THE DAY
THE PHONES
STOPPED

The Computer Crisis—
The What and Why of It,
and How We Can Beat It

BY LEONARD LEE

DONALD I. FINE, INC.
NEW YORK

Library of Congress Cataloging-in-Publication Data
Lee, Leonard, 1960–
    The day the phones stopped / by Leonard Lee.
        p. cm.
    ISBN 1-55611-264-5
    1. Computer software—Reliability. 2. Computers—Reliability.
    I. Title.
    QA76.76.R44L44   1991
    005—dc20   90-56060
                                                              CIP

Manufactured in the United States of America

10 9 8 7 6 5 4 3 2 1

Designed by Irving Perkins Associates

To Mom and Dad

CONTENTS

ACKNOWLEDGMENTS

THERE are many people who were instrumental in assembling this book. Several experts spent a great deal of time answering a barrage of questions to help me grasp the nature of computer software, and for that, I'm most appreciative.

In particular, I'd like to thank those who helped guide me through a maze of government documents to find the hearings, studies, and investigations I was searching for. Several people patiently assisted me, even when I wasn't precisely sure what it was I was looking for. In addition to people mentioned in the book, special thanks to the hard-working staff members of several congressional committees: Deborah Ward and Joe McDonald, House Judiciary; Mike Nelson, Senate Energy; Dave Narcowitz, House Science, Space, and Technology; Shelly Wooly, Senate Armed Services; Bob Weiner, House Government Operations.

In the corporate world, I owe much to Jean Castle of Lou Harris & Associates, Betty Mock of the Incoming Call Institute, and especially to Mike Pruyn and Gary Morgenstern of AT&T, who in their openness and willingness to answer all questions, I believe, have set a new standard for corporate communications in dealing with corporate disasters. In the military, gratitude goes out to Lt.

Col. Johnny Marshall of the U.S. Army Safety Center at Fort Rucker, Alabama, Lt. Dane Joy in the Navy Press Office, Susan Hanson of the Defense Science Board, and Jill Geiger at the Judge Advocate General's Office.

Joyce Pedley and the staff of the Canadian Ministry of Transport were a tremendous source of patient assistance in hunting down past air accidents.

Sharyn Cooper, *Computerworld*; David Nather, *Dallas Morning News*; Ray Formanek, Jr., and David Pace of the Associated Press; and Micki Siegel, I thank you for lending a hand to a fellow journalist.

Special thanks go to family and friends who provided support, and a willingness to listen to a near-constant drone of ideas. I greatly appreciate the efforts of my agents, Virginia See and Jack Caravela, and my editors, Andrew Zack and Adam Levison, who helped whip my rambling thoughts into coherent prose.

And most of all, to Karl Jacoby for his valuable advice, and to my sister Marie for her endless faith and encouragement.

INTRODUCTION

THESE days, it seems you can hardly turn around without getting run over by a computer. They are everywhere. In our microwave ovens, buried deep inside the engines of our cars, controlling our traffic lights, running our switchboards. As computers increase in both number and complexity, the opportunities for mistakes grow. At some time or other, we've probably all been a victim of a computer error. A phone call that got misrouted by a PBX switchboard. A computer workstation at the office that accidentally erased a key piece of data. A new car that stalled because of a flaw in a computer chip. A flight to a vacation destination or a key business meeting that was delayed because of a malfunction in an air traffic control computer. Some of these mistakes are minor. An error on a credit card billing can be corrected with a phone call. Sometimes. Often mistakes are caused by human error, but more often they are due to a flaw in the computer software.

Software, or computer programming, is simply the set of instructions that tell the computer what to do. Without software programming, a computer is simply a very expensive paperweight. It needs to be given directions. Given those, it will follow its instructions very precisely.

1

This book makes numerous references to the term "lines of code." A single line of software code might read:

WRITE (6,*) FIN02 CRG

If you were to give a friend directions on how to drive to your house,

TURN LEFT AT THE THIRD STOPLIGHT
or
GO NORTH ON BOURBON STREET

might be a line of code in those directions.

Now imagine you had to define exactly what you mean by "stoplight." And that you had to explain what "north" means. In addition you have to explain how to operate a car.

Now imagine you have to write out in the minutest such detail, directions on how to get from your friend's home in Seattle to your doorstep in Miami. Such a set of instructions could contain a million lines of code. A year later, you laboriously finish the task. You review it a couple of times to make sure you got it right and that you haven't left out any important steps.

You mail it off.

A week later you get a call from your friend.

From Canada.

He is lost.

You now have to pull out your instructions and go through all one million lines of code one-by-one until you find out exactly where your instructions went wrong. Lots of luck.

In 1945, a team of top Navy scientists was developing one of the world's first electronic computers. Suddenly, in the middle of a calculation, the computer ground to a halt, dead. Engineers pored over every wire and every inch of the massive machine to trace the malady. Finally, one of the technicians discovered the cause of the problem. Buried deep inside its electronic innards, crushed between two electric relays, lay the body of a moth.

One of the scientists remarked, "Well, there's the problem. It seems we have a bug in the computer." The term stuck. A computer "bug" now means an error in software programming. The term is not as innocent as it appears to be.

Even the tiniest of flaws in software can have monumental consequences. It can cause air traffic control computers to mix up airline flights circling in the skies over an airport. Or turn a medical radiation therapy device into a lethal weapon, administering fatal doses of radiation. Or it can put the nation's long distance phone lines on hold.

The irony is that in a software error, the computer is in perfect working order. It is not a mechanical malfunction. It is a failure to correctly instruct the computer what to do.

Software errors have existed as long as there have been computers. But software has traditionally been put through a rigorous testing or "debugging" process that is designed to find and correct bugs before the software is released for public use.

But as new uses are found for technology, software is growing in complexity beyond its ability to be properly tested. And as computer software grows more complex, the opportunities for errors grow. Flaws go undetected until they suddenly cause computer systems to crash, with results ranging anywhere from the merely inconvenient to the terribly tragic. They can cost a business millions of dollars. Or cost taxpayers hundreds of millions. Or kill 290 people in the wink of an eye. All because of a flaw in tiny, invisible lines of code.

Worse, the software error does not have to be a mistake. Even when computer software correctly does what it is supposed to do, it can cause problems. It can encounter situations unanticipated by whoever programmed it. Software will only do whatever it is told to do.

And as software becomes more sophisticated, and we come to rely on it more and more, the problems and their consequences become ever greater.

One congressional investigator warned, "The AT&T failure is just a shot across the bow."

The worst is yet to come.

Chapter One

DANGEROUS MEDICINE

KATY Yarbrough was hoping this would be the last she'd have to see of the inside of a hospital for a while. Although she was recovering nicely, she had grown weary of the constant regimen of doctors, tests, more tests, and therapy. It had been a long month. She hoped her doctors were right, and she was nearing the end of a painful though mercifully brief traumatic episode. Just weeks earlier, she had undergone a lumpectomy to remove a small tumor.

Katy was looking forward to resuming her usual active schedule. The forever-robust sixty-one-year-old Georgia native was an avid swimmer, golfer, and skier. The tumor had been a nuisance. Now that it was over, her doctors assured her she would be as good as new. It was June, and Katy looked forward to building back the strength in her swing out on the links. It could take her all summer to get her handicap back down.

But first, there would have to be more treatment. She had already made it through the rigors of chemotherapy, where even the resulting loss of hair failed to dim her spirit. Today Katy was at the Kennestone Regional Oncology Center in Marietta, Georgia for some minor touch-up radiation treatment. It was quite a routine procedure; the Kennestone Center conducted similar sessions dozens of times each day.

A nurse directed Katy onto the table underneath the radiation therapy machine. The massive Therac-25 linear accelerator loomed imposingly overhead. Kennestone officials were proud of their Therac-25 installation, one of fewer than a dozen in the country. The machine, barely two years in operation, offered much more flexible doses of radiation than the old cobalt radiation machines it replaced.

Sophisticated computer controls allowed physicians to precisely meter the amount and type of radiation to be delivered. The -25 designation was an indication of the immense power of the machine: up to twenty-five million electron volts could be directed at the patient. The unit could generate searing X-rays, capable of destroying tumors deep within a patient's body, yet without damaging surrounding tissue. Alternately, it could offer low-power electron beams that would gently remove tumors near the surface of the skin.

It was the low-power electron beam mode that was selected for Katy's treatment this day. As she lay back on the table, the nurse instructed Katy to lie still and keep her eyes closed. The nurse walked out of the room; from a computer console in the adjoining room, a technician activated the Therac-25. Katy had already been through several of these daily sessions. They were nearly becoming routine. She knew there would be no more sensation than one would get from a chest X-ray.

Within the space of a millisecond, Katy knew she was in trouble. She let out an anguished cry of pain. She could feel a burning cylinder she described as "just solid heat" searing through her shoulder. She said it felt as if someone had rammed a hot pipe through her body. Though her eyes had been tightly closed, she had seen a blinding flash of blue light.

The nurse and the technician came running back into the room. Katy was trembling and in great pain as they gently helped her down from the table. She was quickly taken to an examining room, where a doctor examined a fresh wound roughly the diameter of a quarter that had burned into her shoulder. The doctor seemed surprisingly unconcerned. He told her it was just a mild skin reaction.

Katy described her ordeal while in the radiation therapy room, saying she had been burned by the linear accelerator. The doctor replied that was impossible. Atomic Energy of Canada, Ltd., the manufacturer of the Therac-25, had a reputation as a world leader in radiation therapy machines. The Therac-25's advanced computer software held numerous safety interlocks that would shut the machine down in the event of a radiation overdose.

Katy was sent home. She simply hoped that, as the doctor had told her, it would get better. But as the weeks progressed, her condition worsened. A charred, blackened hole developed on the front of her shoulder. She was in constant pain, which steadily grew sharper. It became more and more difficult for her to use her left arm. Katy Yarbrough never went back to the Kennestone Oncology Center. But other doctors she saw were equally puzzled by her condition. A radiation burn does not usually appear right away. Under a typical low-dosage radiation burn, the damage is slow to develop. One physician examined it and proclaimed it to be a minor skin irritation. He prescribed a skin balm for it.

But Katy's suffering grew more intense. Because of the constant pain, she could not sleep for more than thirty minutes at a stretch, and even then, only because of sheer exhaustion. Finally, her daughter-in-law, a nurse, saw the extent to which she was suffering, and reasoned Katy's pain was far too severe for a minor skin irritation. She was taken to a hospital and given Demerol, which offered her the first relief from the pain in weeks. A more thorough examination at last revealed the wound for what it really was: a radiation burn. The damage she had suffered in less than a second was frightening. The beam of electrons from the Therac-25 had burned a hole almost completely through her shoulder, destroying nearly all the tissue in its path. The brachial-plexus, the bundle of nerves which controls the shoulder, had been burned beyond repair. Doctors could not even properly assess the total extent of the damage, because radiation burns have a slowly degenerative effect; it would take weeks, or even months, before the full damage would be seen.

Katy sued the Kennestone Center, arguing the Therac-25 had maimed her. Mary Ellen Griffin, Katy's attorney, says, "Kenne-

stone told us, 'No, that's impossible. That's not what that is.' "
Griffin, who is also a registered nurse, says she knew what the
problem was as soon as she saw Katy Yarbrough. "We were the
first people that even entertained the thought that she was right. We
looked at her and we just said, 'I don't know how they did this,
but this is radiation.' She did not have a through-and-through hole,
but what she had in the front was a hole, and on the back you could
see a red mark, which was the exit area."

Officials at Kennestone steadfastly maintained that was impossible. The Therac-25 enjoyed a perfect safety record. In thousands
of treatments since the machines' introduction in 1983, no one had
ever suffered a radiation burn because of a Therac-25. Kennestone
announced it would fight the lawsuit. They never notified the manufacturer, AECL, since the incident was not judged to have been a
machine problem.

But less than a month later, a second mishap occurred. A woman
was being treated for cervical cancer at the Hamilton Clinic in
Ontario, Canada. A micro-switch failed, improperly setting up the
machine. The woman received an overdose of radiation.

Investigators from the Canadian Department of National Health
and Welfare concluded that the problem was a mechanical one.
Since the computerized safety interlocks in the Therac-25 had failed
to stop the machine from firing, AECL installed new software interlocks on all Therac-25s.

Nearly one full year passed. Kennestone and other cancer treatment facilities in the United States and Canada continued using the
Therac-25.

When Ray Cox entered the East Texas Cancer Center in Tyler,
Texas, he hoped it would be for his final radiation treatment. Cox
was about to receive a ninth dose of radiation to treat remnants of
a shoulder tumor that surgeons had successfully removed weeks
earlier.

Cox was placed face down on the table beneath the Therac-25.
The two technicians left the room to operate the machine's controls.
They could neither see nor hear Cox. The intercom was not working

that day, nor was a video monitor connecting the radiation room with the nearby technician's console, only a few feet away in an adjoining room. When the machine was activated, Cox experienced a pain he described as feeling like an electric shock shooting through his shoulder. He saw a bright flash of light, and heard a frying sound. Seconds later, a second burst of energy struck him in the neck, causing his body to spasm.

Cox jumped off the table and yelled for help, pounding frantically on the door. Technicians rushed to his aid. Although they did not know it at the time, Cox had just received two massive doses of radiation.

As soon as Ray Cox's treatment session was halted, doctors at ETCC notified AECL, the manufacturer of the Therac-25. AECL engineers in Canada had technicians at ETCC run some standard diagnostic tests on the Therac. The computer's internal diagnostic software revealed nothing amiss with the machine. ETCC officials decided to place the Therac-25 back into service, and patient treatment sessions continued.

The next day, Cox began spitting up blood. He was rushed to the emergency room of Good Shepherd Hospital in nearby Longview. Doctors examining him found he had developed a lesion on his back. Doctors admitted him, and diagnosed a new condition: Horner's Syndrome. Cox had suffered nerve damage. He lost much of his sweat function; his eyelids drooped, and his pupils were dilated. He would spend the next five months confined to a bed in a hospital. He would later become paralyzed and lapse into a coma. Ray Cox would die in September. The cause of death according to court documents filed by his attorneys: respiratory problems brought on by radiation exposure.

At the news of Cox's illness, ETCC's Therac-25 was immediately pulled out of service again. Within two days, a team of technicians from AECL arrived to inspect the unit. The machine was run through its paces at various radiation settings, and performed perfectly. AECL officials said they would have the machine checked to make sure it was properly grounded. They thought perhaps Cox had somehow suffered an electrical shock. No electrical problems were found. The machine was placed back in service.

Two weeks passed before Verdon Kidd came into ETCC to have a small cancer near his right ear treated. The sixty-six-year-old former bus driver took his place on the treatment table; his head was placed in the path of the electron beam.

In the next room, a technician, following standard instructions, tapped B on the computer keyboard, instructing the Therac-25 to fire. The machine fired, but just as quickly disengaged. The display readout flashed "MALFUNCTION 54." By this time, in the treatment room, Verdon Kidd was already moaning for help. He told doctors it felt as if his face were on fire.

This time, the alert technician had spotted the sequence of events that somehow triggered the fatal malfunction. The Therac-25 offers two modes of operation, depending on the treatment needed: X-ray and electron beam. In X-ray mode, an intense beam of electrons is generated, using up to the machine's full-rated power of twenty-five million electron volts. The powerful stream of raw energy passes through a tungsten target which converts it into therapeutic X-rays. The target absorbs the powerful electrons, allowing only X-rays to pass through. This process is extremely inefficient, and requires a tremendous amount of energy in the beam. When the weaker electron beam mode is selected, the target is automatically retracted, and the energy level of the beam is drastically reduced.

The technician recalled the exact sequence of events. She had first accidentally selected the X-ray mode. Using the up-arrow key on the keyboard, she quickly typed over the error and re-selected electron beam mode. The computer display reassuringly changed to show "ELECTRON BEAM."

But inside the body of the linear accelerator, a software error was causing drastic changes to take place. The instant the operator picked X-ray mode, the accelerator selected an appropriately high-intensity energy level. When the technician made the correction and changed it to electron beam mode, the particular combination and speed of keystrokes activated a hidden flaw in the software programming.

The tungsten target which converts electron beams to X-rays swung out of the way, but the machine still remained poised to deliver the stronger radiation level of the X-ray mode setting.

In each case, the machines' safety mechanisms had shut the accelerator down, but not in time to prevent the damage. The interlocks shut the machines down in 3/10ths of a second. But the Therac-25 puts out three hundred pulses each second, each pulse rated at 250 rads of radiation. As Gordon Symonds of the Canadian Department of National Health and Welfare explains, "The machine was misconfigured by the software. What it did was, it cranked it right up to its maximum radiation output rate with essentially nothing in the beam like a target or a beam modifier. The initial safety systems did shut the machine down, but it took three-hundred milliseconds. In that 3/10ths of a second, the patient received about twenty thousand rads, which was lethal in some cases." The dosage was 125 times higher than the average radiation received from an accelerator. The machine had essentially cooked the tissue in the patients' bodies in much the same way as a microwave oven cooks a piece of hamburger.

Through it all, the Therac-25 had remained oblivious to what it was doing with its rampaging radiation. The computer display merely blandly indicated the machine had correctly shut off early, and administered only a fraction of its low-power prescribed dose.

Officials at ETCC immediately notified Atomic Energy of Canada, Ltd., of their horrifying discovery. AECL, in turn, duplicated the error on one of their machines. They swiftly contacted all Therac-25 users in the United States and Canada. The treatment centers were all told to physically and immediately remove the up-arrow key from their keyboards. AECL began a close examination to find the fatal flaw in their computer software.

On May 2, Verdon Kidd died. That day, AECL filed an incident report with the U.S. Food and Drug Administration, which had originally approved the use of the Therac-25 in the United States.

While their investigation was still underway, the Therac-25 struck again. This time the setting was a cancer treatment center in Yakima, Washington.

Although the Therac-25 detected something was amiss, it still failed to prevent the machine from firing when the technician issued the "BEGIN" command. The result was the same as in the previous incidents: the patient received a radiation overdose, although in this case he survived. Somehow, the technician in Yakima had stumbled on to a new path to the same problem.

In November, a team of physicists and engineers from the FDA's Center for Devices and Radiological Health arrived at AECL's manufacturing plant in Canada. They began a thorough investigation of the machine, and in particular the software program that runs it. The FDA wanted to determine how the fatal flaw had been manufactured into the machine.

The Therac-25 error was triggered only in the hands of particularly nimble keyboard operators. The up-arrow key on the computer keyboard was used to edit instructions entered on the keyboard. As the technician entered the specific data about the therapy session—dosage, power rating, beam mode, etc.—the up-arrow key could be used to change any of the settings, or alter a command that had been incorrectly entered.

In this case, quick usage of the key to change "X-RAY" mode into "ELECTRON BEAM" mode triggered a fatal sequence of events inside the machine. As AECL discovered, the particular combination of keystrokes caused the software to fail "to access the appropriate calibration data." The machine would set itself up in a unique configuration, poised to deliver a fatal blast of radiation. Tests by AECL and the FDA found the error would activate only when technicians edited their instructions less than eight seconds after it had been entered into the computer.

The tiny error in the software had laid dormant and undetected for years, waiting for the particular set of circumstances that would cause it to go berserk. Roger Schneider, the director of the FDA's Office of Science and Technology told a House subcommittee investigating the issue of computer software safety that the error would not have been caught, according to the subcommittee staff, "even if the device had been subject to the most stringent pre-market approval."

But, in truth, the Therac-25 had been given far less than rigorous scrutiny before it was approved. The FDA does not itself review the software that goes into computerized medical devices. Like many federal agencies that now must oversee complicated computer-controlled equipment, the FDA simply lacks the expertise, technical staffing, and time to extensively test each piece of equipment for *safety*.

The difficulty arises in that *reliability* and *safety* are two separate concepts. *Reliability* is a determination of whether the program will perform the function it is designed to carry out. Repeated testing will eventually bring a program to the point where it performs its function reliably enough to be released for general use.

Safety means ensuring the software does not malfunction in a manner that endangers people. But software safety becomes a difficult standard to measure because programmers can never imagine the countless different ways in which a program can fail, or the many different ways a program can be used. Therein lies the insidious nature of software errors. As we shall see, time and time again, software errors often go undetected until a particular set of peculiar and unforeseen circumstances causes the errors to manifest themselves.

The Congressional Office of Technology Assessment, which advises Congress on science issues, has looked into the issue for years. A recent COTA report notes that government agencies are helpless when it comes to safeguarding medical software, noting that "there are no standard measures or ways of assessing software safety."

Our reliance on computers adds to the problem by creating a false expectation of reliability and safety—an unwillingness to believe the devices can fail, that they can become unsafe.

Attorney Mary Ellen Griffin points out how officials at the Kennestone Oncology Center in Georgia never notified AECL or the FDA following the first incident involving Katy Yarbrough in 1985. "What they said is 'it's impossible. It couldn't happen,' because they were so willing to believe that their interlocks and computer software were exact and precise."

In truth, the Therac-25 had been poorly designed, in a manner that may have encouraged hazardous use. Therapists at the Kennestone Center report the machine routinely gave out up to forty error messages a day on its computer display.

The error messages were difficult to interpret because the machine would only display a number, instead of a written message explaining the error. Most were minor, and did not present any danger to the patients. Such error displays could be quickly cancelled by pressing the *P* key on the computer keyboard, meaning "PROCEED." Therapists got into the habit of doing this almost every time an error appeared.

When Ray Cox received his overdose, the display flashed "MALFUNCTION 54," indicating either a mild overdose or mild underdose. The therapist quickly cancelled the error and fired the machine twice more.

This type of accident would have been nearly impossible on competing linear accelerators, which rely less heavily on software-driven controls. Ironically, AECL's previous linear accelerator, the Therac-20, contained electrical circuits that would have prevented such a malfunction. Griffin says, "You just cannot rely on a computer software program and a computer driving these machines without being absolutely certain that the safety interlocks are going to kick in."

It is worth noting that Siemens, a competitor of AECL, has a similar radiation therapy machine called the Digital Mevatron. Although it is computer-controlled, Siemens built into it numerous mechanical safety features that would have prevented a similar type of accident.

A device called an electron injector monitors the level of energy being fed in to create the radiation beam. If the energy is too high for the type of treatment being called for, the treatment is cancelled before the beam is even finished being generated.

An additional safety device measures the output of the machine. If the beam is measured to be too powerful, it shuts off the machine. Siemens designers insisted on using mechanical safety features, rather than relying on computers.

Siemens technicians say computer software cannot operate fast enough to safety control these machines. Siemens project manager, John Hughes, points out, "When you have a beam concentrated in that intensity, you've already done the damage before the system could even react. Even in today's technology, there's no way you can write a software program that's fast enough."

Officials at Siemens say use of computers is of great value to radiation therapy machines. The greater flexibility of radiation dosages reduces the risk of the beam damaging healthy tissue that surrounds cancer tumors. It also permits one machine to perform treatments that used to require dozens of different machines.

But for all critical safety points where patients could be endangered, computer software safety features in the Mevatron are backed up by mechanical safety features.

More worrisome is the fact that the Therac-25 is only one small facet of a growing problem facing modern medicine and its patients. Computer technology is quickly moving in on all aspects of health care.

Software-controlled medical devices are growing at a prodigious rate. Nearly $11 billion worth of medical electronics are produced each year: everything from dialysis monitors to anesthesia breathing machines. Frank Samuel, Jr., president of the Health Industry Manufacturers Association, freely acknowledges the tremendous growth in medical computer technology. "The impact of computers on medical care," he says in *Science News* (vol. 133, March 12, 1988), "is the most significant factor that we have to face. Health care will change more dramatically in the next ten years because of software-driven products than for any other single cause."

And as the number of products grows, so do the problems. In recent years, several medical computer products have been found to contain potentially fatal software bugs, such as

- a pacemaker that sometimes produced no output
- an insulin infusion pump that could generate potentially fatal air embolisms in the bloodstream
- a cardiac output monitor that overestimated cardiac output

- a blood analyzer that displayed incorrect values because it added, rather than subtracted numbers
- a patient monitoring system that mixed up blood pressure, heart rate, and other vital sign readouts between different patients
- a ventilator that caused the flow of oxygen to the patient to drop without warning

In one horrendous incident, the FDA discovered that faulty blood bank software had destroyed some donor records. As a result, blood tainted with the AIDS virus and hepatitis was accidentally released for patient transfusions.

In 1989, the FDA issued thirty-four recalls of medical devices because of software problems. That figure was triple the number of recalls in 1984. In most of the cases, "Something was incorporated into the software during the development of that software, some defect," explains the FDA's Fred Hooten. Most of the time, he says, "the defects weren't found until the device was distributed."

But as medical devices become more sophisticated, the FDA is in danger of falling behind in its efforts to keep up with medical technology. FDA officials admit they face serious shortcomings in regulating computer-controlled medical devices. At most, it can review whether the manufacturer is following good industry standards in developing its software. It must rely on the manufacturer to test and debug its own programs. As John Dellinger of Advanced Technology Labs, Inc., a Washington State-based medical technology firm, says, "There have been instances of [federal] investigators making technical calls regarding design that they apparently are not qualified to make."

Both federal agencies and manufacturers are struggling to come to grips with the problem of ensuring safety. But the FDA is searching for a fail-safe means to evaluate software and rate its potential for disasters. As the House Science, Space, and Technology Committee sadly notes, "Such tools do not now exist. At the present level of understanding in software engineering, federal agencies cannot be assured the software they oversee and use is correct.

There is no infallible method whereby a regulatory agency can assure that the software embedded in a system will not cause death or injury.''

There is also resistance within the industry to increased federal regulation. Dellinger fears his industry will fall victim to the same problems that befell military and space technology companies when NASA and the Pentagon imposed regulations on software development. Cumbersome federal procurement regulations required companies to hand over the rights to any new technologies they invented while developing government software. ''The chilling adverse effect of over-regulation on software development,'' he says, ''has resulted in software, generally, for these applications that is five to ten years behind the leading edge technology and still isn't perfect.''

The issue is extremely complex, and goes beyond the mere technical difficulties of software development; for it is a regulatory question as well. Software developers have consistently resisted any efforts to force program writers to be licensed. A person with no medical knowledge can write a program for a linear accelerator, or any other critical medical device.

Gordon Symonds of the Canadian Department of National Health and Welfare notes the disparity. ''If an engineer designs a bridge and it falls down and kills someone, then he can have his license revoked by the professional engineering society and he wouldn't be building any more bridges. Maybe the same thing might one day come to pass for software developers.''

Self-policing by the software community may be a way to develop more rigid standards for software. Robert Britain, as manager of the Diagnostic Imaging and Therapy Systems Division of the National Electrical Manufacturers Association, has been spearheading an effort to develop industry guidelines for software-controlled medical devices. As he notes, ''there are many software development techniques that produce safe, high-quality software.'' NEMA suggests that computer medical devices include written descriptions of the design and development methods used, discussions of potential hazards, and quality assurance procedures.

But even Britain cites the critical shortcoming in software development, notably that "complete review of software testing is often impractical."

Prior to the Therac-25 incidents, AECL enjoyed an excellent safety record. It had also developed software safety systems for Canadian nuclear reactors. Gordon Symonds points up the inherent difficulties of software. "They [AECL] have extreme corporate depth in the field of software safety. They have very, very qualified people to do their software safety analysis. So even with a company like that, there can be problems. It's a very difficult issue."

The Therac-25 has since been corrected. The errant software has been rewritten and reincorporated into every unit in use. Still, nervous treatment centers demanded stringent safeguards against software errors. Suspicious of computer-controlled equipment, users insisted on fully mechanical safety devices that would operate independently of the software. On July 29, 1987, more than a year after the death of Verdon Kidd, the FDA approved a corrective plan. More than twenty modifications were made to existing Therac-25s, which by that time were no longer being manufactured.

All Therac-25s now contain what's called a one-pulse shutdown system. If there is an irregularity in the programming, the linear accelerator shuts down in just three milliseconds, less than 1/300th of a second. In that time, the machine could generate only one pulse of radiation, administering a dose far less than lethal.

But the problem of software in medical devices continues to grow. The relentless push of technology constantly creates more numerous and more complicated uses for computer software. And large, complicated software programs inevitably contain errors. As the House Committee on Science, Space, and Technology ominously notes, "For the Food and Drug Administration, the Therac-25 is a harbinger of problems that will only grow in the future."

Katy Yarbrough never again got to practice her golf swing. Because of her radiation burn, she completely lost the use of her left arm and was forced to keep it in a sling. Her left breast had to be surgically removed. She had scars from a skin graft running from just above her waist to over her shoulder. Her lawsuit against the

Kennestone Regional Oncology Center was eventually settled out of court, and she received compensation for her horrendous disfigurement.

Tragically, Katy died in an auto accident five years after her encounter with the runaway radiation machine. Until her death, not a day went by when she did not experience pain. Just a few months before Katy died, her attorney said, "I saw this woman in tears for months because this pain was intractable. They cut the nerves all over the site so that she does not actually feel the burn anymore. She still has pain, but it's controlled." Mary Ellen Griffin pauses and thinks. In her many years as a nurse and an attorney she has seen it all. "It's sad. I've never seen a person suffer the way Katy did."

But even as the medical field struggles to find answers to the problem of software errors, it turns out that medicine isn't the only place where software errors can kill.

Chapter Two

THE UNFRIENDLY SKIES

TAKE heart, weary travellers. Not every software error works against you. In 1988, a Delta Airlines computer accidentally gave hundreds of travelers free tickets they didn't earn. A glitch resulted in seven hundred people receiving frequent flier bonuses that they were not entitled to. Bill Berry, a Delta spokesman, explained to the Associated Press as best he could, "There was a computer error." Enough said.

After the error was discovered, Delta told the passengers it would honor their free bonuses. Said Berry, "We invited those individuals to use those bonuses with our compliments."

Another two thousand passengers, on the other hand, ended up not being credited with frequent flier miles for flights they had actually taken. Delta quickly rectified that situation as well.

But, as veteran travellers know, most software problems involving airlines rarely turn out as well. More typical were the problems Continental Airlines encountered in the never-ending search for airline automation.

On May 22, 1988, Continental Airlines opened, with much fanfare, its new Terminal C at Newark, New Jersey. Continental modestly billed it as "the most convenient, most modern, most comfortable, most exciting airline terminal in the world.'' The quarter-billion dollar edifice was said to be the most technologically advanced building of its type. The interior had been designed with pearl-white tile floors and soft lighting to try and create what the company termed a "comfortable atmosphere" for travellers arriving from and departing to their planes. One wonders why Continental didn't simply dub it "The Eighth Wonder of the World." The terminal, which serviced only Continental flights, was larger than all of La Guardia Airport.

The first sign things were not going well came as passengers and even flight attendants became lost in the cavernous terminal. Small wonder. At 875,000 square feet, the facility made passengers sort amongst 41 departure and arrival gates and 103 different check-in and ticketing points.

Worse, Continental had decided to use Terminal C to venture into the previously uncharted waters of computerized baggage handling. Each piece of luggage was given a tag with an electronic bar code. As the baggage was unloaded off the planes, computerized scanners would read the codes and direct the bags through a series of conveyor belts to the proper luggage carousel. But bags were quickly misdirected, the result being a forty-five minute wait for baggage, when under manual control it took only twenty-five. Some passengers complained that the wait for their baggage was longer than their flights.

Continental officials took it all in stride, saying the bugs would be worked out in time. They were, although it took a bit longer than they would have liked. The problem was the optical reader, a laser that read the bar-code stickers that had been placed on the sides of baggage. When each piece of luggage was checked in, it was given a sticker with a printed code, similar to the bar codes found on grocery items in supermarkets.

Technicians were confused; after all, the system had performed flawlessly in test runs. But in a real-world setting, the ticket agents applied the bar-code stickers in ways and places the optical scanner couldn't decipher. Continental spokesman Dave Messing admits,

"It was the volume. We had a huge number of bags on a system that had only been tested in an unrealistic setting putting test bags through it."

As you'll see, unrealistic testing and unanticipated situations are a common theme in the world of software problems. In this instance, fortunately, it resulted in nothing more serious than lost luggage.

Messing claims the system has been fine-tuned, and that Terminal C now has the world's finest and fastest computerized baggage-handling system.

NOWHERE is the growing use of software more controversial than it is in aviation. Software errors here have potentially catastrophic consequences. Even the tiniest of errors can endanger hundreds of people at a time when they put jumbo jets on collision courses, or cause military planes and jetliners to simply fall from the sky.

Up in the air, jetliners are being transformed into flying computers. Norman Augustine, chairman of the Martin Marietta Corporation, in his autobiography *Augustine's Laws* (Viking Penguin, 1986) joked, "After the year 2015, there will be no airplane crashes. There will be no takeoffs either, because electronics will occupy one hundred percent of every airplane's weight."

In truth, what has happened has been the exact opposite. In the never-ending search for ways to make planes more fuel efficient, manufactures have shaved thousands of pounds off of airframes by replacing bulky hydraulic systems, control cables, and analog instruments with lightweight electronics.

Manufacturers have done the impossible. They have found something that makes jetliners cheaper to operate, more reliable, and yet weighs nothing. Software. Invisible lines of code that operate computers. It provides an irresistibly simple way for airlines to save on fuel, maintenance, and crew costs all at the same time.

Sophisticated fuel management computers are making jetliners far less expensive to operate. The fuel savings that result from the computers' precise metering of fuel to suit varied flying conditions has greatly contributed to reduced expenses for the airlines in an era of rising fuel prices, helping to hold down the cost of flying.

The replacement of hydraulic and mechanical controls with computer electronics also greatly simplifies maintenance. Computer software doesn't break, wear out, or need oiling.

Computers greatly reduce the workload of flight crews. Engine monitoring and navigation computers now mean that a two-person cockpit crew can do the work formerly requiring a three-person crew, offering considerable cost savings to the airlines. Even the largest airliner in the world, the Boeing 747-400, the newest version of the 747, requires only a pilot and a co-pilot. And even then, the individual workload is less than it was on previous 747s.

To do this, however, requires tremendously complicated software. The first generation of jetliners introduced in the 1950s contained virtually no computer software. But modern airliners are operated by sophisticated computers that contain millions of lines of code. And this opens the door to a new hazard of modern travel: software error. Airline passengers and crews are finding themselves victims of situations never anticipated by computer software designers.

In July 1983, Air Canada Flight 143, bearing seventy-two passengers and crew members, nearly became the tragic victim of both aircraft and air traffic control computerization. Cruising over Red Lake, Ontario, the flight suddenly ran out of fuel. Capt. Bob Pearson, a twenty-six-year veteran of Air Canada, found himself commanding a powerless machine. Pearson's first thought, *How could this have happened?*

The Boeing 767 was one of the newest planes in the fleet, having been delivered only three months earlier. The new-generation plane featured what pilots call "a glass cockpit." Gone were the traditional mechanical instruments that once cluttered airplane cockpits. All vital information was now provided by `six cathode-ray displays, similar to TV screens.

In addition, the 767 uses sophisticated electronics to continuously monitor its fuel supply. A tiny microprocessor, itself a small computer, sits immediately behind the cockpit underneath the cabin floor. The processor continually calculates the quantity, weight, and even the temperature of the jet fuel in each of the 767's three fuel tanks. The system controls fueling valves during refueling operations, and even contains its own troubleshooting programs to check for problems. Gordon Handberg, program manger for Honeywell, which developed the 767 processor, noted proudly to Canadian investigators after the incident that the system has "the ability to sustain a failure somewhere in the system and continue to operate."

To ensure reliability, the system uses two independent data channels to relay the information from the fuel tanks to the computer fuel quantity processor. The system is designed so that if one of the data channels fails, the other automatically takes over, providing an automatic backup.

But one of the circuits aboard the Air Canada 767 contained a tiny mechanical error. A bad solder joint in one electrical connection gave one of the two data channels only partial power. The processor knew how to handle a complete loss of power in one channel, but did not know what to do under a partial power loss. The software in the system had not been designed to compensate for this particular error. Instead of switching to its backup system, the entire computerized fuel monitoring system simply shut down. The system had stopped working on the last flight prior to 143. The cockpit crew now had no indication of how much fuel they were carrying.

Before leaving Ottawa, Captain Pearson and mechanics had decided they could repair the problem later. They had the tanks filled manually, taking precautions to ensure the proper amount of fuel was loaded onboard.

But pilots and fuelers quickly became confused, because while the 767 measures the weight of the fuel in metric kilograms, Canadian fueling trucks dispense it by volume by the liter. Due to a miscalculation, the plane took off with twenty-six thousand pounds less fuel than it needed to reach its destination.

It was just after 8:00 P.M. over Red Lake when Flight 143's left engine stopped operating. Captain Pearson and his crew were still struggling to find the cause when the twin-engine jet's other powerplant also went quiet. The loss of power in any airplane is alarming; in a computerized plane it is catastrophic. All the electronic displays in the cockpit suddenly went blank. All electrical power comes from generators in the engines, or from an auxiliary power unit in the tail, which also depends on fuel.

Suddenly, Pearson had only the most rudimentary instruments at his disposal. Boeing had incorporated standby analog instruments in case of problems: an altimeter, an old-fashioned magnetic compass, an airspeed indicator, and an artificial horizon. Pearson was now flying his jetliner with fewer instruments than Charles Lindbergh had at his disposal in 1927.

Actually, he was even worse off. The magnetic compass was located high above the windshield between the two pilots. Pearson found he had to lean over to see it. Even then, neither man could read it accurately because of its awkward angle. It was impossible to use. Pearson decided to judge the accuracy of his course by looking at the layer of clouds outside the windshield.

Underneath the floor of the cockpit, in the electronics bay of the airplane, a standby twenty-eight volt battery kicked in. It supplied just enough power to run the essentials needed to keep the plane in the air. All the flight controls still functioned. Cabin pressure was maintained. A single radio would maintain contact with the outside world. Pearson now had thirty minutes in which to land the plane before the battery would run out.

Emergency lights flickered on in the cockpit. Together with the last dim orange rays of the evening, they cast an eerie glow in the cockpit. Pearson was able to read a quick reference emergency guide. He swiftly followed the instructions to extend the plane's emergency ram air turbine. The "RAT" as it's colloquially known, was extended below the fuselage. Air from the plane's slipstream rushed through the turbine, generating enough hydraulic pressure to enable Pearson to gain minimal control of the airplane. Pearson now had electrical and hydraulic power, but just barely.

The next order of business was to establish the plane's optimum glide profile. Captain Pearson obviously wanted to stretch the plane's range to the limit, to enhance its chances of staying aloft long enough to reach a landing strip. Every airplane design has a particular airspeed that will allow the plane to maximize its gliding distance without power. But that information was nowhere to be found in any of Boeing's flight manuals or Air Canada's training manuals. The plane also lacked a standby vertical airspeed indicator to help the crew judge their rate of descent. *Boeing engineers had considered it impossible that a 767 could ever lose power in both engines at once*. Good fortune was riding with the passengers and crew of Air Canada Flight 143 though. Bob Pearson happened to be an experienced glider pilot, who had spent hundreds of hours piloting powerless planes.

By consulting with air traffic controllers who were tracking Flight 143 on radar, Pearson was able to guess roughly how far the plane would be able to glide.

What was essential was to bring the plane down as quickly as possible. Without any of its computerized navigational instruments, Air Canada 143 would now depend on air traffic controllers to guide it to a safe landing. Here, too, Flight 143 nearly fell victim to a problem in software design.

Controller Ronald Hewitt was keeping watch on Flight 143 from deep within the bowels of the Winnipeg Air Traffic Control Center, housed in a dark, windowless room at the Winnipeg International Airport. When 143 lost power, it was located some sixty-five miles to the east of Winnipeg. Hewitt was in the process of directing the plane by radio to its destination in Edmonton when Flight 143 suddenly vanished from his radar screen.

Radar, developed by British scientists in the late 1930s, operates on a simple principle. Radio waves are beamed from an antenna. The metal surfaces on airplanes at altitude reflect the beams, which are then returned to the source. Through timing the difference between the transmission and the time the echo returns, radar can determine how far away an object is.

But modern air traffic control radar operates on a much more advanced level. Radar cannot differentiate between targets: all targets look alike. Controllers cannot at a glance tell one flight from another, or commercial traffic from military or private planes. So when radar signals strike a jetliner, it activates an electronic transmitter called a transponder. The transponder sends out a *coded* signal to the ground radar. The code appears next to the radar image of the plane, allowing the controller to identify each plane under his control.

When 143 lost power, its transponder failed, along with most other electrical equipment on the plane. Ronald Hewitt's computer-driven radar, which like most air traffic control radar in the world, recognized only coded transponder signals, no longer recognized the plane, and its image disappeared from the scope. Its software was deliberately programmed to ignore transponderless airplanes.

But luck was once again riding with 143. Hewitt still had at his disposal an old primary radar system. Of the eight radar stations he could monitor, only Winnipeg still offered the primitive, outdated primary system that depended only on the ground transmitter and receiver to find airplanes. Flight 143 was lucky to still be within range of Winnipeg radar's coverage. At most other air traffic control centers in the world, 143 would have been invisible to their computerized radar. Hewitt hastily fumbled with his controls. The image of 143 reappeared on his screen. Hewitt could now guide it to what he hoped would be a safe landing.

Flight 143 was descending too quickly to make it all the way to Winnipeg. Pearson contemplated landing on a highway, or possibly ditching in Lake Winnipeg. Controllers scrambled to find alternate airports. They had to measure 143's position with cardboard rulers, because designers had decided the controllers would not need sophisticated computerized distance-measuring equipment built into their air traffic control software. Flight 143 was losing altitude quickly. The plane was dropping out of the sky at a rate of twenty-five hundred feet per minute. From its initial altitude of thirty-five thousand feet, the 767 had lost considerable altitude. Pearson now had only scant minutes to find a landing spot. Controllers noticed

143 was only twelve miles from an abandoned Royal Canadian Air Force strip at Gimli, Manitoba. Controller Len Daczko gave 143 a heading that would allow it to reach Gimli before the plane ran out of altitude.

Without hydraulic pressure from the engines, flaps could not be extended on the wings, and the plane would land at a much higher-than-normal speed. Pearson hoped he could stop the 767 in time on the short seven-thousand-foot-long Gimli runway. As the plane approached the field, Pearson ordered co-pilot Maurice Quintal to lower the landing gear. Quintal pushed down on the gear handle. The two men waited for the reassuring *swoosh* as the gear extended below the fuselage. Nothing happened. Without full electrical power or hydraulics, the gear could not be lowered.

Quintal frantically flipped through the 767 Quick Reference Handbook in the Aircraft Operating Manual. He looked through ''landing gear'' and ''hydraulics'' and could find no reference on how to manually lower the gear. Finally, he flipped the alternate gear extension switch on the front cockpit panel. The switch released the pins holding the undercarriage in place, and the landing gear dropped out of its wells by its own weight. But the rush of the airstream prevented the nose gear from locking into position.

At 8:38 P.M. local time, seventeen long, harrowing minutes after the 767 lost power, Flight 143 touched down on runway 32 Left. Two tires blew out on the main landing gear, and the nose gear collapsed. The giant jetliner slid along the runway, scraping its nose and engine nacelles on the tarmac. People who were gathered on the runway for a sports car rally scrambled to safety as the jet slid to a stop. Only minor injuries resulted.

In the ensuing investigation, Canada's Ministry of Transport laid most of the blame on the pilots and the mechanics for failing to carry out proper fuel calculations, and for dispatching a plane with an inoperative fuel monitoring computer. But the Board of Inquiry also laid part of the blame on Air Canada for not properly training pilots and mechanics on how to calculate fuel loads in the event of fuel quantity processor failure. The company had simply not adequately anticipated that such an occurrence could ever happen.

The Board called the Gimli accident "an indictment of the training programs set up for this new aircraft." The report said the training for this complex aircraft was "geared to the optimistic forecasts of the manufacturers and developed in contemplation of ideal situations . . . it does not cope with the real world."

The report recommended the plane be equipped with additional standby instruments and a better emergency power supply that could power a transponder. The same argument could probably be made for all "glass cockpit" jetliners, although it is worth noting that newer jetliners are being made with wind-driven generators and larger batteries to supply standby power.

As the relentless march of technology continues, there are inevitable tradeoffs among comfort, efficiency, and safety. With each new generation of modern jetliners, computers play a more dominant role, and the software that controls them becomes more complicated. Software now controls everything from fuel management systems to galley ovens. But as computer software comes to dominate today's modern jetliners, does it make them safer or more dangerous? The answer is both. It makes planes safer in some respects, but also creates new hazards. The best example of this may be the Airbus A320.

February 22, 1987. Toulouse, France. Officials of Airbus Industrie watch proudly as their newest airplane, the A320 is about to take wing for the first time. Company officials and government dignitaries from throughout Europe nod their heads approvingly and admire the sleek jetliner as it slowly rolls to the runway threshold at the Aerospatiale test facility at Toulouse-Blagnac airport. At the controls, Airbus chief test pilot and vice president of Flight Division Pierre Baud is ready to usher in a new generation of jet travel: one that relies on computer software to an extent never attempted before. Airbus officials are both nervous and excited. Any maiden flight is nerve-wracking, but at stake today is no less than the future of their company. For them, the A320 represents their best chance yet to break the dominance of American giants Boeing and McDonnell Douglas, which together effectively rule the entire industry.

For Airbus, even getting to this point has been a rocky, uneven road. Formed in 1970, Airbus Industrie is the first concerted attempt by European manufacturers in a decade to break into the lucrative and technologically important field of commercial jetliners. Airbus is a massive, though sometimes unwieldy consortium, representing the best efforts of its partners, British Aerospace Corporation, France's Aerospatiale, CASA of Spain, West Germany's MBB, and their four respective governments.

Despite massive government subsidies and some of the best aerospace talent in Europe, Airbus's first two offerings were only modest successes. The A300, and its slightly smaller cousin the A310, sold over four hundred units through the mid-1980s, but were unable to crack the massive U.S. airliner market, save for a token sale of thirty A300s to Eastern Airlines in 1977. In 1987, Airbus was still losing money on every jetliner it sold. Britain, France, West Germany, and Spain continued to pour millions of dollars in government subsidies into the consortium. Airbus Industrie desperately needed a breakthrough, some sort of a leap in technology that would set it apart from other airliner manufacturers.

Along came the A320 and Airbus officials were justifiably proud of their latest offering. For the first time, they had produced a jetliner that was clearly superior to its American competitors, and computers and advanced software were credited with that.

Airbus officials smiled as Pierre Baud smoothly advanced the throttles, and the A320 swiftly accelerated down the runway, not with the characteristic roar of jet engines, but instead with more of a gentle but persistent hum. The A320 would be one of the quietest jetliners in the world while offering a huge operating cost savings by incorporating software on a scale never seen before.

In an era of rising fuel prices and increasing worldwide competition, the A320, a small jetliner carrying around 150 passengers, could offer airlines fuel savings of up to forty percent, compared with older, less advanced jets such as the Boeing 727.

The A320 would be the world's first jetliner to extensively use what are known as fly-by-wire controls. Other modern jetliners use a host of computers for everything from navigation to fuel management. But the A320 would be the first jetliner to be almost completely software-controlled.

Conventional jetliners use control systems whose origins date back to the first Wright Flyer in 1903. When a pilot pulls on his/her controls, long metal cables that stretch to the control surfaces (rudders, ailerons, and elevators) on the wings and in the tail of the airplane. Hydraulic systems move the surfaces, much as power steering in an automobile makes steering a heavy car an easy task.

But in a fly-by-wire system, the pilot uses a video-game-like joystick. The control senses how much pressure the pilot is applying and in what direction. Those instructions are then relayed to a computer. The software in the flight computers is programmed to convert the pilot's inputs into actions. The software then commands hydraulic actuators to move the various sections of the wing and tail.

Engineers built huge redundancies into the system to ensure safety. Seven separate computers control the various flight systems aboard the A320. This is designed to ensure more-than-adequate backups in case one computer should fail. It also protects against multiple failures. Utilizing a software structure referred to by programmers as "a paranoid democracy," the computers constantly monitor each other. If there is a disagreement, the computers determine by consensus which computer is "wrong," and then "vote" to ignore it.

Fly-by-wire theoretically offers several advantages over conventional control systems. Engineers estimate the use of fly-by-wire took about seven hundred pounds off the A320 by eliminating heavy control cables and cutting down on hydraulic lines. Seven hundred pounds may not sound terribly substantial when dealing with a 160,000 pound airplane, but the reduced weight can save airlines millions of dollars in reduced fuel expenses. The sophisticated software programming also constantly adjusts the aircraft's controls without the input of the pilot, trimming control surfaces so that the plane slides through the air with a minimum of air drag. The software makes the A320 one of the most fuel-efficient jetliners in the skies.

Proponents argued fly-by-wire would also enhance safety. Bernard Ziegler, Airbus Industrie's senior vice president of engineering told the assembled media he considered fly-by-wire technology to be the single biggest improvement in flight safety in history. The flight control software, he explained, will override a pilot's commands if the computer feels the pilot is requesting a maneuver which would endanger the airplane. The software determines when the pilots have let the plane slip into a dangerous attitude, or the plane is descending too quickly, and automatically compensates.

But, it clearly was the economics that drew airlines to the A320. Airbus officials had finally pulled off the major coup they had been waiting eighteen years for: the A320 was irresistible. Even before the first one had rolled off the assembly line, airlines around the world had lined up and placed orders and options for more than four hundred, a record for any new commercial jetliner. More importantly, Northwest and Pan Am, two major U.S. airlines, had finally been lured onto the Airbus bandwagon. The prestigious carriers agreed to buy up to 150 of the A320s, a deal worth more than $6 billion.

Once airborne, pilot Pierre Baud put the prototype through a full range of flight maneuvers: high speed, slow speed, steep turns, steep climbs, emergency descents. The seven flight control computers accepted his commands and flew the plane with precision. Baud noted only two minor problems: a slight yawing or side-to-side motion during some maneuvers, and a slow engine response when the throttles were advanced. With previous prototype airplanes, expensive structural changes would have had to be made to the airframe. But now, the problems could easily be fixed by simply altering the computer software. An engineer would come aboard later with a portable computer and reprogram the software. That was the beauty of fly-by-wire.

Three hours after it departed, the prototype A320 returned to Toulouse-Blagnac airport. Baud made two low passes over the field to get a feel for how the plane handled on landing approaches. He then circled around a third time, and touched the plane down smoothly on the main runway. As the plane taxied up to its hangar, Airbus officials could not help but feel that this would be the plane that would finally bring the consortium the success it had been waiting for for so long.

But as Airbus officials and workers watched the first A320 roll to a stop, they could scarcely imagine that just sixteen months later, one of the very first A320s would go down in flames, and ignite a controversy that continues to this day over the extensive use of software in airplanes.

JUNE 26, 1988, dawned sunny and bright in the skies over eastern France. This was welcome news for Capt. Michel Asseline. Sight-seeing was to be an important part of the day's agenda. As Air France's chief instructor for the new A320s, Asseline could look forward to a day away from the routine of flight training, and a chance to do some low-level flying in the beautiful eastern section of France where the majestic Alps would be within easy view.

The forty-four year old Asseline was an experienced veteran, with over ten thousand hours at the controls. Even though A320s were relatively new to the fleet, he had already logged over 130 hours flying them. His co-pilot for today, forty-five-year old Pierre Mazieres was equally experienced, and well acquainted with the A320.

Today's flight plan called for something out of the ordinary: an appearance at an air show. Asseline and Mazieres would thrill spectators gathered at the Habsheim airport near the town of Mulhouse with a series of low-level fly-bys. For the two veteran pilots, it would be a welcome chance to put the A320 through its paces and demonstrate both their flying skills and the capabilities of the new plane. For Air France and Airbus Industrie, it would be easy public relations.

Early that Sunday morning, the two pilots took off from Charles de Gaulle Airport in Paris for Basel-Mulhouse airport, near the Swiss border. Two special charter flights there had been booked through the airline's affiliate company Air Charter. The passengers would be flown over the air show at Habsheim, taken on a short hop to see the spectacular nearby Mont Blanc, and then returned to Basel-Mulhouse by mid-afternoon. It would be a short flight, perhaps forty-five minutes in duration. One hundred thirty passengers would be aboard the first flight. Some of the passengers were journalists, others flying buffs who had each paid a tidy sum for the flight. Twenty of the passengers were particularly excited about the trip. They had earned the ride by winning a local newspaper contest. For some of the winners, it was their first trip in an airplane.

The Airbus was in splendid condition. It was only the sixth A320 to be delivered to the airlines, and had begun service just four days earlier.

Just after noon, Air France Charter Flight 296Q left the Basel terminal. It would be the first of two round-trips for the crew. As the gleaming white Airbus taxied to the runway, Captain Asseline briefed his co-pilot on the plan for the air show fly-by.

Asseline was anxious to show off the capabilities of the A320, of which he was an ardent supporter. Transcripts taken from the cockpit voice recorder revealed Asseline's confidence in both the airplane and the maneuver he was about to attempt. ''I've done it twenty times,'' he boasted to Mazieres as the Airbus taxied. Asseline was half-right. Both pilots in fact had performed the maneuver several times in training, but only at high altitude or in simulators. But neither had ever performed it close to the ground or at an air show.

The plan for the spectacular fly-over was a simple one. The crew would make two passes over the Habsheim airfield. The first one would be particularly eye-catching. The jetliner would pass over the airfield at a height of just one hundred feet above the ground. The plane would fly along the runway with its landing gear extended and the nose of the plane raised at a very sharp angle. In fact, the

Airbus would be at its slowest possible speed, with its nose held aloft at the maximum angle permitted by the fly-by-wire controls and their complex software programming. To spectators on the ground, the airplane would seem to hang in the air at an impossibly steep angle and low speed.

Air France pilots loved this maneuver, which they felt displayed the impressive margin of safety the computerized fly-by-wire controls offered. The maneuver had been performed more than a dozen times previously at air shows in France and abroad by other pilots, and had never failed to impress the crowds.

It took the plane mere minutes to approach the Habsheim airport, where the crowds had gathered for the air show. At a range of ten miles from the airport, the pilots spotted the field and set the plane up for its initial pass. Asseline approached the field and throttled back the engines, slowing the Airbus down to a flying speed of about 170 miles an hour. The FADEC (Full Authority Digital Engine Controls) engine computers responded to his commands, and reduced power to the engines. His plan called for him to lower the aircraft to a height of about one hundred feet above the ground. After passing over the runway, he would then advance the throttles, adding power to the engines, climb away from the field, circle around, and perform another pass.

"We are coming into view," Mazieres radioed the Habsheim tower. "Going in for the low-altitude, low-speed fly-over."

The controller in the Habsheim tower acknowledged the Airbus.

The A320 passed over the runway. Spectators pointed and snapped pictures as the plane drifted casually along the airstrip, the Airbus's white fuselage porpoised up in a nose-high attitude. Its CFM56 fanjet engines issued their characteristic insistent low thrum.

Spectators anticipated the plane pulling up and going around the field. But as they watched in disbelieving horror, the Airbus flew past the end of the runway, and settled, almost gently, into a grove of trees. A huge fireball erupted nine hundred yards past the runway's end.

An eyewitness, Bernard Labalette, himself a pilot, described the scene to the French news agency Agence France-Presse: "It was about twenty meters off the ground with its flaps and landing gear down when it clipped some trees at the end of the runway and crashed in the forest."

The Airbus slashed a path through the forest 150 yards long and 50 yards wide. One passenger told Agence France-Presse, "We were passing over the Habsheim runway when I heard trees rubbing under the cabin. Then things happened very quickly. The plane scraped the ground and we slammed to a halt. We had a sensation of everything standing still, then the fire began in the front of the aircraft." Before the fuselage came to a halt, the right wing broke away. Fuel spilled from the wing tanks and ignited.

Rescue workers rushed to the scene. Flight attendants heroically stayed with the burning plane and managed to evacuate nearly the entire cabin, using the emergency exit chutes. The fuselage was mostly intact, but the fire was quickly spreading. Rescue crews struggled to free between ten and twenty-five passengers trapped in the wreckage. Many passengers were dazed after hitting their heads on the seats in front of them. Further complicating the situation, the cabin lights went out, and the Airbus's computerized emergency lighting failed to activate. Passengers began panicking and started pushing their way through the heavy smoke and flames.

Despite these difficulties, the evacuation by the flight crew, which would later win the praise of authorities for its efficiency, was credited with saving many lives. But three people died, two of them children. Thirty-six others had been injured.

The fire continued to burn. When it was finally extinguished, there was little left of what had been a brand-new aircraft, save part of the left wing, and the tail with its distinctive red-white-and-blue insignia.

Even today, the debate still rages in France as to whether the use of fly-by-wire controls saved lives, or itself was the cause of the crash. The incident eventually led to a lawsuit between the pilots and the members of the French transport ministry. That lawsuit remains unresolved.

Even as rescue workers assisted the survivors, investigators immediately tried to piece together what had happened. Both Asseline and co-pilot Mazieres survived, although both were clearly dazed and in shock from the impact. Asseline quickly pointed the finger of blame at the software in the Airbus's flight control computers. Asseline mumbled to a rescue worker, "I wanted to increase power, but the aircraft didn't respond."

The official French accident report noted the crew's contention that the airplane had failed to respond properly: "the behavior of the engines was questioned by the crew immediately after the accident: they declared that, after initiating go-around, engine thrust was not achieved."

The French pilots' union, in an apparent attempt to protect one of their own, said they supported Asseline's view of events, and blamed the A320's sophisticated fly-by-wire controls. Airbus officials quickly rose to the plane's defense. But Air France, British Airways, and Air Inter, a French domestic carrier, the only airlines flying the A320 at the time, grounded their planes.

One day later, the airlines lifted the order, and resumed A320 flights. British Airways issued a statement saying, "We are fully satisfied with the safety and integrity of the craft."

Asseline maintained he had passed over the runway at the planned altitude of one hundred feet. But when he advanced the throttles, he claimed, the computers had ignored his commands, and failed to respond.

French safety investigators quickly determined the fly-by-wire software had functioned perfectly. In fact, Asseline had inadvertently disconnected some of the computer modes, which ironically would not have permitted the fly-by, but instead would have automatically assigned take-off power to the engines, and forced the plane to a higher altitude. The A320 software has a wind shear protection mode, known as alpha floor. If the plane is held in a nose-high attitude of greater than fifteen degrees, the computer automatically adds power to prevent the plane from descending. As they were approaching Habsheim, the two pilots discussed their intention to override the alpha floor protection if it activated during their fly-by. What the two men failed to remember was that the alpha floor protection mode is programmed to disengage at altitudes below one hundred feet, allowing the pilot to raise the nose for landing.

When the airplane approached the runway, Asseline pulled the throttles back to the flight idle position, where the two jet engines would produce only a small amount of thrust. The A320 descended rapidly as the airport drew closer. As prescribed for the maneuver, Asseline pulled the nose of the Airbus up fifteen degrees above the horizon to slow it down. As planned, the plane drifted lazily along the runway, about one hundred feet above the tarmac.

"One hundred feet," called out the Airbus's computer voice altimeter over the cockpit loudspeaker.

"Okay, you're at one hundred feet there," Mazieres warned. "Watch, watch."

But the airplane continued to drift lower. "Fifty feet."

"Okay. I'm okay there," Asseline said reassuringly.

"Forty feet." The computer altimeter continued to call out the plane's steadily decreasing altitude. The computer voice recited the altitude as programmed in its flat, mechanical tones, without any sense of urgency.

The end of the runway was coming up quickly. The Airbus was still descending and slowing down.

"Watch out for the pylons ahead, eh," Mazieres warned. "See them?"

"Yeah, yeah, don't worry," Asseline replied.

Finally, Asseline slammed the throttles forward.

"Thirty feet."

Five seconds later, the Airbus impacted the trees.

That the plane had been that low to begin with was inexcusable. French air safety regulations require a minimum altitude of 170 feet for such overflights. A standing Air France internal note clearly states that air show fly-bys conducted with the plane in landing configuration must be performed at a minimum height of one hundred feet above the ground. In addition, that minimum is only for airfields capable of landing the particular airliner. Since the Habsheim runway was a mere seven hundred yards long and therefore incapable of landing an A320, the fly-by should, according to Air France procedures, have been conducted at an even higher altitude. The Airbus had drifted to an altitude of less than forty feet before the pilots shoved the throttles forward, advancing the throttles to their full forward stop at the full thrust setting.

Despite Captain Asseline's repeated contention the A320 did not respond with power, the software in the FADEC engine control computers worked as programmed. Investigators determined "once the throttles were advanced, the engines responded in the normal time, and perhaps even a little faster than normal." Asseline had simply waited too long with the aircraft in a precarious condition: at low altitude and slow airspeed, with the airplane sinking, the plane slowing down precipitously, and the engines at idle thrust. Normally, whether computer-controlled or not, an airplane takes about eight seconds to go from idle power to full thrust. Jet engines have a notorious lag in the time it takes them to power or spool up. Captain Asseline applied full power less than five seconds before impact, when the airplane was barely thirty feet off the ground. The flight recorders clearly show the engines responded and managed to reach ninety-one percent power. But by that time, the engines had begun ingesting tree branches, and seconds later trees began striking the fuselage.

Daniel Tenenbaum, director of the Direction Generale de l'Aviation, the French civil aviation ministry, concluded, "When the pilot advanced the throttles, the thrust increase was normal, but it apparently was made too late." In a final, desperate attempt to gain altitude, Asseline pitched the plane up to a nose-high attitude of twenty degrees. But lacking sufficient power, the plane clipped forty-foot high trees some two hundred feet beyond the end of the runway, and then settled in a grove of trees beyond.

Not only was the fly-by-wire control software exonerated, but some experts contend it prevented the crash from being more serious than it was. They say the computer controls helped keep the wings level as it crashed, which caused the plane to impact the ground in a straight line, rather than tumbling and splitting the cabin into sections, which would have thrown passengers out of the wreckage. The French investigation noted, "Its path in the forest was practically straight and always in line."

In addition, the plane maintained a nose-high attitude, impacting with its tail, instead of striking nose-first. One eyewitness quoted in the *New York Times* (vol. 137, June 27, 1988) said, "The plane did not go into a nosedive. It belly-flopped onto the trees." This cushioning effect of the trees is credited with lessening the forces of impact and saving lives.

Investigators say the fly-by-wire controls allowed the crew to control the aircraft until the moment it contacted the ground. They wrote, "the trajectory followed by the aircraft was consistent with the commands given by the pilots through the [fly-by-wire] flight control system." The software in the seven flight control computers had worked perfectly. In the preliminary accident report, lead investigator Claude Bechet, himself an Air France pilot, concluded, "No evidence of technical problems with the aircraft or its equipment have been found."

The investigation did note, however, the plane's emergency lighting system failed because of a design error in the computer software. That error has since been corrected.

The Habsheim accident did not cost Airbus Industrie any orders. Six months later, the Federal Aviation Administration awarded an operational certification to the A320, making it the first fly-by-wire jetliner to be certified in the United States for commercial passenger operation.

In the end, the French DGAC Transport Ministry cited the cause of the Habsheim crash as "pilot error." The two pilots and the French pilots' union disputed the claim. Then two years later, a second A320 crashed under mysteriously similar circumstances.

Indian Airlines Flight 605 was in the final stages of its flight to Bangalore, India on February 14, 1990. Valentine's Day. It was nearly a full flight, with 139 passengers aboard. The A320 was new to Indian Airlines' fleet, having just been introduced weeks earlier. C. A. Fernandez was flying for the first time from the left-hand captain's seat of an A320. If the check pilot on his right, S. S. Gopujkar, checked him off, Fernandez would make the much sought-after transition from the rank of co-pilot to pilot. But first, there was the flight to complete.

The stage from Bombay south to Bangalore had so far been uneventful. The weather as they approached Bangalore Airport was almost perfect: skies nearly clear, with excellent visibility and only a light crosswind. But as the red-and-white Airbus turned on to final approach for runway 9, Fernandez inadvertently selected a computer flight mode called "open descent mode." The selection commanded the engines to throttle back to idle thrust, producing virtually no forward power. Open descent mode is meant to be used only when descending from a flight's cruising altitude. It is not designed to be used on final approach to landing. The jet quickly began to sink below the proper approach path.

Fernandez pulled back on his control joystick, raising the nose in an attempt to arrest the increased rate of descent. Neither he nor his check pilot noticed that because of the low power setting, the A320's airspeed had dropped to a dangerously low level.

The plane was now descending rapidly. Clearly alarmed as the ground rushed towards the plane, Fernandez hauled back on the joystick to its full backward stop. "One hundred feet," called out the computer altimeter.

"Sink rate. Sink rate." The flight computers' aural warning system activated, issuing a warning in a computer-generated voice. Fernandez swiftly slammed the throttle levers all the way forward to take-off power.

"Fifty feet."

It was too late.

The A320 impacted heavily on a nearby golf course, seven hundred meters short of the runway. The plane rebounded clumsily into the air and crashed into an embankment, finally coming to rest next to a brick security wall built around the airport.

The right fuel tank ruptured, igniting flames that engulfed the fuselage. Rescue efforts were hampered by a horrendous lack of preparation on the part of Indian aviation authorities. The airport had no emergency plan. There was no radio or telephone link between the control tower and the fire station. Fire and rescue vehicles were not equipped with keys for the single locked gate at the end of the field where the crash occurred. One rescue truck tragicomically attempted to spray foam from the other side of the fence, but could barely reach any of the wreckage. One fire truck quickly became bogged down on the poorly maintained service road, and had to be pushed to the crash site. Another had to proceed at a walking pace.

It took forty minutes for the flames to be extinguished. Ninety-eight people, including the two pilots, died. In the wake of the accident, the Indian Civil Aviation Authority grounded the airline's fourteen A320s. The ban was lifted five months later. As in the case at Habsheim, the investigators blamed the crew for the crash. Fernandez had inadvertently selected an altitude setting in the flight director computer that resulted in the accidental activation of open descent mode throttle settings. The crew had not monitored their speed on approach, and the pilots may have ignored low altitude and sink rate warnings from the flight computers.

But many pilots say the similarities between the Habsheim and Bangalore incidents point out that the problem lies not with the pilots, but with the interface between man and machine. That the pilots are not so much flying the plane as they are merely operating it through the use of a computer. In the A320's case, the software programs in question operate in such a complex manner, with literally dozens of flight modes, overrides, and automatic safety features, so as to be literally too advanced for the task.

Pierre Gille, a 737 pilot for Air France, is also president of the French National Union of Airline Pilots and Flight Engineers. In the wake of the two accidents, he has become an outspoken critic of the marriage of man and machine in the new generation of airliners. "If the Habsheim and Bangalore crashes are both attributed to pilot error," he told *Flight International* magazine (vol. 137, February 28, 1990), "this undoubtedly derives from poor understanding between the machine and the pilot, and here the aircraft builders have to do something."

Gille contended the meshing of pilots and computers in airplanes is fine, "except when the aircraft's behavior produces a surprise to the crew." He conceded there is no evidence of a technical fault in the two crashes.

Airbus Industrie officials, meanwhile, agree with the notion that pilots and flight computers have not forged a smooth union aboard the A320. But there the two sides split apart. The company maintains the problem is the pilots failing to adapt to the automation, rather than acknowledging the need for the software to work smoothly with the humans.

Airbus admits the advanced level of automation in the A320 can lead to crew overconfidence. Bernard Ziegler, Airbus Industrie's vice president of engineering told *Aviation Week & Space Technology* magazine (vol. 132, April 30, 1990), "The crew can begin to believe that anything can be done with the airplane. This attitude could lead to dangerous situations because the physical laws of flight still apply. We [have] warned airlines that crew overconfidence in the A320 was a contributing factor in the accidents, and this overconfidence must be countered."

In investigating the Air France crash at Habsheim, the French Ministry of Transport likewise noted that the computerized safety features of fly-by-wire can induce overconfidence in pilots. "The training given to the pilots," noted the investigation commission, "emphasized all the protections from which the A320 benefits with respect to its lift, which could have given them the justified feeling of increased safety. In particular, the demonstration of the activation of the safety features and protection of this aircraft may lead one to consider flight approaching . . . the limitations [of the aircraft]."

The French investigation commission noted that, in the case of the Habsheim crash, the crew eagerly performed a maneuver that they probably would not have attempted on a conventionally controlled airplane that did not offer computer protection. "During training on previous generation aircraft, pilots are made aware of the dangers of flight at low speed, and the basic rule, which consists of observing a margin with respect to a minimum speed, is driven home. The pilots would therefore probably not have considered such a maneuver . . . However, for the flight of the accident, the crew did not hesitate to plan a low-height fly-over at a speed lower than the normal minimum operating speed."

Pilots may be assuming that the computers will always protect them from disaster. But the concerns over the Airbus A320 extend beyond difficulties of crew overconfidence. The controversy involves the very fly-by-wire design of the A320 and also of future aircraft on the drawing boards: the idea of replacing hydraulic and mechanical controls with software-driven flight computers.

The A320 continues to be an incredible commercial success. Air carriers around the globe have ordered more than eight hundred A320s and slightly larger A321s, sporting the colors of several major airlines, including Northwest, Air France, British Airways, Air Canada, and Lufthansa. Northwest, America's fourth-largest airline, is flying more than a dozen A320s, and will have more than one hundred by 1995. Phoenix-based America West Airlines, the nation's ninth-largest, is also using A320s, and plans to buy or lease at least 74, possibly as many as 118, total.

Airbus Industrie is currently producing ten A320s a month, and the plane is expected to be one of the mainstays of commercial aviation through the early years of the next century.

Airbus will incorporate the same technology into its next generation of planes, the A330 and A340, which will be flown by several airlines, including Northwest and Continental. Airlines have ordered more than four hundred A330s and A340s. The success of its new generation of airplanes has allowed Airbus Industrie to finally achieve a long-sought goal: the consortium in 1990 announced it had achieved its first-ever profit in Airbus Industrie's twenty-year existence.

Now other airplane manufacturers are jumping on the fly-by-wire bandwagon. Boeing, the world's largest maker of airliners, is moving into this computerized technology with its 777 jetliner. United will be among the major airlines flying Boeing's new widebody, which will begin commercial service in 1995. The new jumbo jet, barely smaller than the 747, will have a fly-by-wire system using computers to transfer the pilot's commands into action. It marks Boeing's first extensive use of such technology, although it also uses fly-by-wire to control engines and other systems on some of its other planes.

As fly-by-wire proliferation increases, concerns and fears mount. There are four major concerns with fly-by-wire systems. Authority limitations, systems failure, software error, and electrical interference.

Many pilots say they are nervous about the idea of a plane that has a mind of its own and will ignore some commands. The A320 software programming will not permit a command that could overstress the airframe, or would be, in pilots' parlance, "outside the flight envelope." The idea is to prevent the pilot from carrying out maneuvers that could be potentially dangerous. If the pilot requests a maneuver the software judges to be dangerous, such as raising the nose too high, the software will literally "take over" the controls, and force the nose down and add power. If the pilot tries to bank the airplane too quickly, the computers will go ahead and roll the airplane, but only at a certain rate that it determines to be safe.

Many pilots are understandably skeptical about software that will not permit them to carry out certain maneuvers, even if needed to save the airplane. It assumes the programmers have fully thought out every conceivable dangerous situation that could arise. Harold Ewing is a veteran 747 pilot with over 13,500 flying hours. As he points out, "For day-to-day operations, that's fine. But there are instances such as wind shear where I think the system is an accident waiting to happen." In drastic instances such as wind shear or severe downdrafts while on approach to landing, "you may have to achieve flight loads and attitudes that these computers are specifically designed to reject.''

Other airplane manufacturers have decided not to follow Airbus's lead. Boeing's 777 fly-by-wire computers will never reject a pilot's command. The computers will warn the pilot that a potential maneuver is unsafe, but the pilot will always have the option of overriding the computer; it will never prevent the pilot from carrying out any maneuver. As Boeing spokesman Dick Schleh explains, ''If you apply sufficient force on the controls you can push through those limits. You definitely know when you're doing it, because there's a certain amount of tension on the controls. It takes some real conscious physical effort, but it can be done.''

Boeing engineers feel this feature means a greater margin of safety in those instances where the pilot might need to push the plane to its limits or even beyond. ''If the airplane gets into a critical situation, and you need to use the full design capability of the airplane, you would be able to do that,'' Schleh says. ''And in that real critical situation, that may be the margin that can save the aircraft.''

Dr. Joseph Tymczyszyn, manager of flight deck research for the Boeing Corporation, points to a number of airliners that were saved from crashing only because the pilots used extreme maneuvers that were outside the flight envelope, or the designed capabilities of the plane. A flight envelope limitation mode in a fly-by-wire computer would prevent a pilot from carrying out a similar maneuver. Just the same, Tymczyszyn conceded in the *New York Times* (vol. 138, December 12, 1988), ''There have [also] been accidents in which the pilot flew [beyond the capabilities of the plane] and got himself into big trouble.''

McDonnell Douglas senior test pilot John Miller argues, though, that pilots should always be able to override computers. ''We have to balance the fact,'' he told *Flight International* (vol. 137, May 30, 1990), ''that automation can and will fail; or that the pilot may need to override the system because he may have a situation that the system has not been set up to cope with.'' Although McDonnell Douglas's latest jetliner, the MD-11, is not fly-by-wire, it does feature extensive use of computers. But, pilots can override any and all computer commands. Unlike the Airbus A320, in the MD-11, the pilot, and not the software engineer, remains the final authority in deciding how the airplane is to be flown.

Placing limitations on what pilots can or cannot do in flying their planes may be an unwelcome intrusion, replacing the skills of the pilots with the rigid programmed responses of computers.

A second worry with fly-by-wire controls is the possibility of the loss of all computer systems. Although complete systems failure in fly-by-wire planes is remote, it is not impossible. In March 1988, the first A320 delivered to Air France was taking off from Paris on a demonstration flight with Premier Jacques Chirac aboard. Four of the computers in the controls failed when a transformer malfunctioned. The flight continued without incident. Air France officials say the remaining computers functioned perfectly, and there was no danger. Airbus claims the odds of a complete computer systems failure are about the same as the airplane losing a wing.

Several different power sources are used for the various on-board computers. In the event of a loss of all electric generators, a ram air turbine provides standby electrical power. If that fails, there are two batteries on board.

In addition, the plane can be landed even after the loss of all computers. An electric elevator trim system provides some degree of control separate from the fly-by-wire controls. In test flights, Airbus pilots have managed to land the plane using only this backup, although airline pilots are not required to practice this maneuver.

But that backup system has also come under scrutiny. The electric elevator trim has jammed six times, both in test flights and on commercial flights. In October 1990, the FAA ordered Northwest Airlines to conduct emergency safety inspections on its A320s. The FAA sent an order known as an airworthiness directive. The directive noted that failure of the electric trim coupled with a loss of two flight control computers "could result in loss of control of the airplane."

Northwest and Airbus officials dismissed the possibility of both systems failing as remote. The A320 has a basic wire-pulley mechanism running to the tail section controls, allowing a pilot to retain some control even if the fly-by-wire system and the backup system somehow both failed completely. And they note that even after thousands of flights, these standby systems have never had to be used.

Boeing's 777, however, has no backup systems whatsoever. If the fly-by-wire system were to fail completely, the pilot would have no means of controlling the airplane. Boeing engineers feel they have built enough redundancy into the system to ensure that such a calamity could never happen.

Besides being confident in the integrity of the system, Boeing officials feel that designing in mechanical backup controls defeats the whole purpose of building fly-by-wire systems in the airplane in the first place. "What you're trying to do," explains Boeing spokesman Tom Cole, "is get rid of the weight caused by the pulleys and cables that would be the mechanical backup. That's what you want to get off the airplane."

The 777's three flight control computers each use a separate design in order to prevent all three from ever failing at the same time. The plane also offers an impressive combination of eight separate generators, an emergency wind-driven generator, and batteries that the company claims will ensure a continuous power supply under all possible circumstances.

Engineers insist a complete failure of a fly-by-wire system is so remote as to be insignificant. As these new planes accumulate more hours in the sky, only time will prove or disprove that notion.

The major concern with fly-by-wire aircraft is that even if the systems do not fail completely, they could be vulnerable to software errors hidden somewhere in their programming.

The A320, as all passenger planes do, underwent a stringent battery of tests before receiving its airworthiness certification. More than eight hundred flights were conducted, logging more than 1,650 hours of flight time before the first passenger was carried. The Boeing 777 will likewise undergo stringent testing both in the air and on the ground by the Federal Aviation Administration before it carries its first passenger in 1995.

But software engineers are quick to point out that not every conceivable situation can be played out, no matter how rigorous the testing. Mike Hennell, a professor of computational mathematics at the University of Liverpool, says in *Newsweek* (vol. 112, July 11, 1988) simply, "This technology is dangerous."

Annoying bugs have already shown up in the A320's software. Pilots for British Airways and Air France have reported engines mysteriously throttling up or suddenly losing power while on approach to landing. Cockpit instruments have displayed erroneous altitude readouts. Pilots have noted steering problems while taxiing on the ground.

Pierre Gille, the head of the French pilots' union, says in the British *Flight International* magazine (vol. 137, February 28, 1990), "We think the A320 is a wonderful aircraft, but we all think there are manifestly quite a few technical deviations that need to be corrected." Airbus Industrie's senior vice president of engineering, Bernard Ziegler, says the number of such complaints to date has been small.

The critical question is, could a software error ever compromise safety? There are precedents in other fly-by-wire aircraft. Fly-by-wire was originally developed for the military, but it has by no means been perfected there.

When the first prototype F-16 fighter, one of the first fly-by-wire planes, was developed, it contained a programming error that would have caused the plane to flip upside down every time it crossed the equator. Mike Champness, a former computer software manager for the F-16 project, explained how the goof got into the system. "We had a lot of problems with software. It was a prototype airplane they were working on. It was a case of a mistake in one of the ground referencings for latitude, and when you went over the equator the positive/negative sign changed and the airplane would have gone upside down." Fortunately, the error was caught in a computer simulation.

But some accidents in military airplanes have been directly attributable to problems in software programming. One of McDonnell Douglas Corporation's prototype F/A-18 jet fighters fell victim to software problems. During a test flight, prototype number 12 went into a spin. The pilot attempted to maneuver out of the spin, only to discover the fly-by-wire computers were programmed to reject the maneuver he was attempting. The computer ruled the request was unsafe. The aircraft subsequently crashed, although the pilot managed to eject safely. The software was later modified.

Granted, the situation the pilot found himself in was highly unusual and unpredictable. It took subsequent test pilots more than one hundred tries before they could duplicate the events aboard the ill-fated test flight. But this simply highlights how programmers can never anticipate all the possible situations a plane can get into.

The Saab-Scania Gripen jet fighter was already eighteen months late when it took off for its first test flight in December 1988. The project was also a whopping $1 billion over budget. The delays and cost overruns had been caused by problems in developing the software for the Gripen's complex fly-by-wire computer controls.

Officials had big plans for the program. It was hoped the sleek single-seat fighter could handle nearly all of Sweden's fighter needs into the next century. Swedish Air Force officials planned to buy up to four hundred of the fighters. But the complicated software was still proving troublesome. The Swedish Defense Administration required numerous and stringent tests before it approved the software, even for in-flight testing. Government officials wanted to ensure the software was safe. The final approval from the government had come just the night before. After the first hour-long flight, Saab-Scania's chief test pilot told company engineers the plane's controls were too sensitive.

Still, Saab-Scania officials were clearly relieved. Now that the initial test flight had proven successful, they could look forward to accelerating the test program.

Eight weeks later, Saab-Scania pilot Lars Radestrom took the Gripen up for its sixth test flight. It was Radestrom's first flight aboard the Gripen. Although the five previous flights had been largely uneventful, software developers were still grappling with a problem of unwanted side-to-side motions in the plane. Programmers had underestimated the arm strength pilots were using as they manipulated the video-game-like joystick that controlled the fly-by-wire system.

After an hour of putting the plane through a series of high-altitude maneuvers, Radestrom turned the Gripen back towards the Saab-Linkoping test airfield. In the distance, Radestrom could see that a sizeable crowd of spectators and media was gathered at the airfield. This was the Gripen's first public appearance, and the sleek fighter plane would appear on Swedish television news that night.

Radestrom turned the Gripen onto final approach. It was just before noon. The weather was good: clear with just a few light wind gusts. A reduction of power began the plane's descent towards landing. The airplane nosed down slightly. Lars pulled back on the joystick to pull the nose up. Nothing happened. The nose dipped more. Radestrom was alarmed: the airplane was already close to the ground. He pulled back harder. Suddenly, the nose shot upwards like an elevator.

Radestrom pushed forward to compensate. Again, no response. He increased the forward pressure. Nothing. Then the nose just as suddenly pitched downwards again. Lars felt the plane was behaving drunkenly. The plane was delaying his control inputs, failing to respond quickly to his commands. As a result, he was wildly overcompensating.

By this time, it was too late, and the Gripen was too low for him to pull out of the dive. The Gripen landed hard. The left landing gear immediately collapsed. Radestrom gamely hung on to the controls as the Gripen bounced off the tarmac, landing in a skid that threw the plane off the runway. The Gripen spun and began careening through the dirt and sod backwards at high speed. A wingtip dug into the ground and flipped the plane over. The fighter rolled a harrowing five times before finally coming to a stop.

Europe's newest fighter plane came to a rest upside down. Fire crews smothered the plane in fire-suppressing foam, extinguishing a small fire that ignited. Radestrom was trapped in the cockpit, and crews had to lift the inverted plane to get him out. Miraculously, he was still alive, suffering only from shock and a broken elbow.

Saab-Scania officials traced the problem to the fly-by-wire software. It had not been written correctly to control pitch at low speeds. The accident report concluded, ''The pilot's control commands were subjected to such a delay that he was out of phase with the aircraft's motion. This caused uncontrollable pitch oscillations, which resulted in the aircraft hitting the runway in the final phase of landing.'' The particular combination of low speeds on landing approach coupled with Radestrom's control inputs had activated this particular software error.

The error had not been previously discovered, despite extensive pre-flight tests in both computer simulators and Saab-Scania's ''iron bird''—a massive test rig used to check out actual flight control systems. Gripen program manager Tommy Ivarsson was puzzled. ''We never encountered this situation in the simulator,'' he told *Aviation Week & Space Technology* (vol. 130, June 26, 1989), ''The problem that caused the accident was never detected.'' Saab-Scania delayed the project one year in order to properly modify the software coding.

These accidents occurred during the testing phase, before the planes actually went into production. But testing is no guarantee that all the software errors will be found. Former F-16 software manager Mike Champness points out that not every flight scenario can be tested, and not every error can be caught in advance. ''There are so many different logic paths that you can never really test them all. So you try and test as many as you can, in as many weird and standard situations. But that makes it very difficult. Imagine all those other little insidious things you wouldn't normally think of. It's always the unobvious situations that come back to haunt you.''

Dr. Peter Neumann, a researcher at SRI International, a leading computer research firm, agrees. ''There's no way you can guarantee a [fly-by-wire] computer system's going to do the right thing under all possible circumstances. No matter how carefully you've designed it, it's probably not good enough.''

Government officials admit there are no guarantees the fly-by-wire software in the A320 is error-free. Brian Perry of the British Civil Aviation Authority concedes no fly-by-wire software will ever be perfect. He told the *Washington Post* (vol. 112, April 2, 1989), ''It's true that we are not able to establish to a fully verifiable level that the A320 software has no errors. It's not satisfactory, but it's a fact of life.''

When errors manifest themselves in flight, military pilots at least have the option of ejecting from their planes. Passengers aboard commercial planes unfortunately have no such alternative. Airbus Industrie is quick to point out that since the A320s began flying in 1987, none of the computers in the planes has failed due to a software error.

The last major concern with fly-by-wire aircraft is the possibility of electrical interference. Conceivably, it could override the software, causing the airplane's controls to malfunction. Tests have shown that such common electronic devices as hand-held electronic games, radios, and portable computers generate electrical signals that exceed levels permitted for any equipment installed on aircraft. Airbus Industrie's Engineering Support Group conducted exhaustive tests in 1984. Tests with laptop computers, radios, video cameras, and electric razors showed that, even though several did emit strong radio signals, they did not cause any interference with airplane systems. A special committee of the Radio Technical Commission for Aeronautics reached a similar conclusion.

When the Federal Aviation Administration certified the A320 for use in the United States, it subjected the Airbus to the most stringent tests of any airplane ever certified to make sure it was immune to electromagnetic interference. A test A320 was flown through numerous lightning strikes with no demonstrated effect on its controls.

In spite of such rigorous testing, the A320 proved susceptible to outside interference once it began operating in the United States.

In their first year of operating A320s, Northwest pilots reported a series of problems. There were nine instances where suspected computer malfunctions caused A320s either to abort takeoffs, return to the airport after takeoff, or make unscheduled landings. In one incident the pilot was on initial approach to Washington, D.C.'s national airport. He disconnected the computerized autopilot when he felt the plane was descending too steeply.

On another occasion, a pilot aborted takeoff in San Francisco after the navigational heading disappeared from the cockpit displays.

The problems were eventually traced and corrected with a modification to the flight software. Northwest spokesman Doug Miller explains, "At certain airports where there was a certain kind of instrument landing system [being used] there was some kind of an electrical interference with that system and the computers onboard the A320. Airbus Industrie determined that, and they made a software fix, and the difficulty hasn't recurred since then."

Northwest and FAA officials insist the malfunctions were minor. They claim such glitches are not out of the ordinary when breaking in an aircraft, and that the problems never compromised passenger safety.

The pilots' union agrees. Spokesman John Mazor says the number of problems reported "is not at all out of line. It is true with the A320, as it is with any new aircraft. You can go to any particular aircraft, you'll find it has to go through a teething period, and it takes care of the problem."

But modern airliners are clearly susceptible to outside interference, and one can never quite predict from where it will come. In one instance, a maintenance worker was working inside a jetliner when the outflow valves on the plane's cabin pressurization system suddenly closed by themselves. The cause: radio signals from his portable walkie-talkie. In another case, a jetliner's fire alarms were set off by signals coming from a passenger's cellular telephone.

And there are instances where electrical interference can cause fly-by-wire planes to fly with a mind of their own.

March 13, 1985. An Army UH-60 Blackhawk helicopter takes off on a daytime training mission. As the sleek jet-powered helicopter departs a landing zone at Fort Bragg, North Carolina, it carries a crew of four. Also on board are eight paratroopers from the 82nd Airborne Infantry Division. Accompanied by two other Blackhawks, the helicopter quickly slips in behind the other choppers as the trailing member of the aerial formation.

Today's flight is nothing more than a routine training mission. The ship had successfully inserted a platoon of soldiers into a practice area, and was now returning them to base. This Blackhawk is one of 555 similar machines in the Army's inventory. Made by the legendary Sikorsky Company, the chopper is the Army's replacement for the familiar Huey helicopter that served in Vietnam. Considered the most capable and easily maintained troop-carrying helicopter in the world, the Blackhawk is a more-than-adequate replacement for the venerable Huey, and performed well for the Army in deploying troops in the recent invasion of Grenada.

But Blackhawk Chalk Three never returns to base. As eyewitnesses aboard the other two helicopters watch in stunned disbelief, the trailing helicopter suddenly pitches sharply upwards, then pitches sharply downwards, plummeting until it impacts the earth at a near-vertical angle, upside down and facing backwards. The Blackhawk explodes and burns. All twelve servicemen are killed.

Although this is the twenty-second accident since Blackhawks were introduced in 1981, the Army is quick to point out that despite a grounding order a year earlier after two Blackhawks crashed, the helicopter still enjoys an excellent safety record.

Investigators from the Army Safety Center in Alabama arrive and begin investigating the cause of the crash, as they do for every such incident. They detect no signs of any malfunction in the helicopter's flight control systems. They note the pilots appear to have been fighting back against the controls when the craft impacted into the gently rolling woods. The exact cause of the accident remains a mystery.

The Blackhawk is, by and large, a run-of-the-mill aircraft. Its controls are activated by a standard combination of mechanical flight controls and hydraulic assists, with one critical exception. The tail stabilator, the large horizontal control surface at the tail, is fly-by-wire. The stabilator, which combines the functions of the elevator and horizontal stabilizer of an airplane, controls the helicopter's pitch, that is, the up-and-down movement of its nose. It is the only control surface on the helicopter that is computer-controlled. The large tail flap moves up and down independently depending on airspeed, the attitude of the helicopter, and the pilot's control inputs. By doing so, the software programming helps the pilot maintain control when he flies at high speeds low to the ground, a mission profile Army pilots often must fly to deposit troops quickly behind enemy lines.

In April 1987, a Blackhawk is flying across fields twenty miles southwest of Wiesbaden, West Germany. Suddenly, and with no warning, the helicopter noses over. It is pitched downwards at a steep seventy degree angle, and accelerating. The helicopter plunges two thousand feet in barely fifteen seconds, a sickening rate of descent. To the occupants, the craft feels as if it is heading straight down. As the ground rushes towards the Blackhawk, the co-pilot notices the stabilator is jammed in nearly the full "down" position. Although both men struggle against the errant control, the chopper continues to plunge towards the earth.

Just as suddenly as it began, the stabilator moves back to its normal position, and the crew is able to regain control of the helicopter. The shaken pilots slow the Blackhawk down and ease it to a safe landing. Nothing in the cockpit indicates there has been a malfunction.

Upon investigating, startled Army crews discover the cause. The Blackhawk was flying near a large radio broadcasting tower. Engineers determine that electrical impulses from the tower had somehow commanded the controls to throw the helicopter into a steep dive. The stabilator controls have a built-in fault monitoring system that is supposed to detect any computer problems. The electrical interference had overridden the software programming and caused the stabilator to move by itself in a dangerous manner. After six months of testing, the Army ordered that all Blackhawks be retrofitted with increased shielding, and switches that will allow pilots to override unwanted commands.

Although subsequent laboratory testing showed that electromagnetic interference from ordinary sources such as power lines and broadcast towers can cause the software-driven controls to move on their own, the Army still denies such interference has caused any Blackhawks to crash. In a videotape distributed to all UH-60 pilots, Col. William Turner, project manager for the Blackhawk program, assured them that none of the anomalies encountered should critically affect flight safety.

Still, in the laboratory, the Army could never manage to duplicate the severity of the problem experienced by the Blackhawk crew in Europe. Critics charge electromagnetic interference with software-driven controls may have been responsible for five Blackhawk crashes that killed twenty-two servicemen.

In fact, the tests found that radio waves could trigger a complete hydraulic failure, which could lead to loss of control of the aircraft. Army tests also found the UH-60's cockpit warning lights were susceptible to electromagnetic interference. Warning lights would go off at random, or else would fail to go off when they were supposed to in the presence of electromagnetic interference. It was discovered that a walkie-talkie could interfere with the chopper's master caution/warning system. In all, forty out of the Blackhawk's forty-two system were susceptible to electromagnetic interference from radio and radar transmissions.

All Blackhawk crews are now issued special maps that show the several hundred electromagnetic sources around the world that can interfere with Blackhawk controls, and pilots are ordered to stay clear of those areas. It is also worth noting that since the problem with electromagnetic interference was identified, the Army's accident rate for UH-60s had dropped dramatically. And the Navy, whose version of the Blackhawk, the SH-60B Seahawk, has always carried extra shielding against electromagnetic interference, has consistently enjoyed a much lower accident rate for their helicopters.

The point is not that fly-by-wire controls are necessarily more dangerous than conventional controls. Indeed, the safety record of the A320 indicates, to this point at least, that Airbus Industrie's claims may be true, that fly-by-wire's use of software-directed controls in commercial aviation actually makes planes safer. No crash of an A320 has yet been directly attributable to fly-by-wire problems.

For the most part, pilots, once trained to the A320, are satisfied with it. John Mazor, spokesman for the Airline Pilots Association, has spoken with Northwest pilots, and relates, "the pilots are generally enthusiastic about it. It does require some different techniques of flying, but once you've been trained and become familiar with that, I think most pilots are very pleased with the airplane and pretty confident with its capabilities."

Many pilots who were initially skeptical about using fly-by-wire have now become enthusiastic supporters of the technology. But pilots also report it requires completely different training. "You almost have to think about your airplane in a different way," Mazor explains, "than you did with the old mechanical linkages and analog dials. Can this put you in a difficult situation? Yes, it can. But that's true of any airplane."

Still, Mazor expresses a few misgivings about the technology. "Obvious concerns that you have are what happens if something goes wrong. Lurking in the background, of course, is the principle that you can never anticipate everything."

Concerns over computerization extend beyond fly-by-wire airplanes such as the A320 and 777. Even though they are not completely computer-controlled, nearly every other airliner made today features extensive use of computers. The software use in these computers is extensive. At one point, McDonnell Douglas and its subcontractors employed about fifteen hundred software engineers to develop the MD-11's myriad of computer systems. The use of computers can greatly simplify pilots' workloads, allowing them to concentrate on safe flying. The cockpit, instead of being a bewildering nerve center, is a much simplified hub of activity. The MD-11 cockpit contains sixty percent fewer gauges, switches, and indicator lights than that of its predecessor, the DC-10.

In particular, the systems are designed to reduce distractions during emergencies by automating procedures. Dumping fuel, an action often required before emergency landings, used to require the flight engineer to activate twenty-eight separate switches. With computer automation, it only requires a single command. When the plane is flying too slowly, such as in a wind shear condition, flaps and slats are automatically extended on the wings to provide extra lift to keep the plane in the air.

In a sense, fly-by-wire is simply the latest manifestation of the growing trend towards software in all new jetliners. And the advanced technology is presenting us with an entirely new set of safety issues, which to this point neither the software nor aviation industries have a great deal of experience dealing with.

Software can cause problems even when it works perfectly. A major problem is that software often does not properly take into account the human element: how people react to software, and how it affects their decision-making. The vagaries of human behavior and how they interact with machines are often not worked into the design of software where safety is an important factor.

Automation in commercial aviation is meant to reduce the chances of human error. But people will always stumble across the loophole in the software that programmers overlook. Time and again, human error has shown it can manage to trip up even the most fail-safe system.

One of the worst airliner disasters in history may have been caused by such a programming error.

August 31, 1983. Korean Airlines Flight 007 departs Anchorage International Airport en route to Seoul, South Korea. The big 747 lifts off forty minutes behind schedule, a tiny delay for a flight that would last seven hours. Most of the 265 passengers settle in to sleep. The flight had departed at 5:00 A.M. local time, and a long last leg lay ahead. At the controls is Capt. Chun Byung-in, a former Korean Air Force pilot with eleven years of airline experience.

Inexplicably, the plane begins to drift off its intended course almost immediately after takeoff. Unnoticed by its experienced crew, the 747 steadily flies a course to the west of its intended track. In the hours that follow, the plane drifts hundreds of miles off course, passing over sensitive Soviet military installations, first over the Kamchatka peninsula, and then over the Soviet island of Sakhalin.

As the world has come to know well, Korean Airlines 007 ended its flight when it was struck by two missiles fired from a Soviet fighter plane. The fighter pilot was ordered to shoot down the un-armed Boeing 747 when the plane was just minutes away from entering international airspace.

Why the jetliner strayed 365 miles off course will probably forever remain a mystery. But speculation is the crew may have simply made a tragic error in programming their navigational computers, aided by a fatal flaw in its software.

Despite common perceptions, most modern jetliners, particularly on overseas flights, no longer navigate by radio, or by sighting the sun or stars, as they did previously. Radio reception of navigation signals over the oceans is often spotty at best, depending on atmospheric conditions. Celestial navigation is too dependent on clear weather. Particularly on overseas flights, most navigation is performed by a system known as INS, or Inertial Navigation System. It is a sophisticated guidance system that is an outgrowth of the navigation systems developed by NASA for Apollo moon missions. A series of sensitive gyroscopes sense any and all movement of the aircraft. By comparing the motion with the aircraft's previous known position, the INS computer can determine, with pinpoint accuracy, the plane's position anywhere on the globe.

But to perform the calculations, the software needs to be "told" where on the Earth the plane is before it begins its journey. Cockpit crews are given a thick manual that lists the precise latitude and longitude, down to the degree and second, of every departure gate of every airport in the world. The information is loaded into the INS computers prior to departure. Once the gyroscopes have aligned themselves, and the plane is pushed away from the gate, the INS will from that moment on continuously display the plane's position. Coupled with the autopilot, the INS will also direct the plane to its destination, and to any intermediate checkpoints along the route.

The software is designed to be as foolproof as possible. Boeing 747s carry three separate INS computers. Each is independent of the others. Information is entered separately into each one through a carefully mapped-out procedure. The three computers continuously cross-check each other, and if there is any disagreement the cockpit crew is alerted that there is a discrepancy.

Prior to KAL 007's departure from Anchorage the morning of August 31, 1983, airline records show Flight Engineer Kim Eui Dong was the first crew member to enter the 747 cockpit. It would then have been his duty to program the INS, and it is here that some surmise an error was made that led to the destruction of 007.

One popular theory is that Kim inadvertently misentered the aircraft's position on the ground in Anchorage. Instead of entering the plane's location as being 149 degrees West longitude, Kim may have keyed 139 degrees West.

The software should have caused the error to be kicked out. To guard against such errors, the airplane's initial position on the ground in Anchorage had to be entered separately into each of the plane's three INS computers. If Kim accidentally entered incorrect coordinates into INS number 1, and then inserted the correct numbers into unit number 2, the system would have flashed a warning light saying "INS Compare," indicating there was disagreement between the two computers. Kim would then have re-entered the coordinates into INS number 2, and again would receive an error message. In this puzzling situation, Kim would naturally assume the difficulty was with INS number 2: after all, INS number 1 was already loaded. INS units are also notoriously sensitive. The slightest movement of the aircraft on the ramp will cause it to reject its internal coordinates. On this particular model, Kim could cancel the warning by simply pressing the warning light. As one veteran pilot said, "It can happen so easily it makes your blood run cold. Flight crews have done things like that many, many times."

Entering the data into INS number 3 would not trigger a further warning, since that warning mode was now disengaged. It is worth noting that other manufacturers' INS units use different software that would have continued to display warning messages and would not have allowed the plane to take off.

That an INS programming error was to blame for KAL Flight 007 is purely speculation. The answer to the mystery of what happened to 007 lies underneath thirteen thousand feet of water, in an INS computer that rests at the bottom of the Sea of Japan. But theorists note that a ten degree error in longitude in that section of the globe would result in the flight path being approximately three hundred nautical miles off course.

The INS programming error theory gained a good deal of credibility when a programming problem nearly caused two jumbo jets to collide. The incident, which endangered nearly six hundred people, highlighted the problems of the growing use of software in the cockpit.

On July 8, 1987, Continental Airlines Flight 25 was on its way from London to Newark. The giant 747 was cruising leisurely over the North Atlantic at thirty-one thousand feet, having left London's Gatwick Airport three hours earlier. The jet was following a course known as Track *D*, one of five assigned air corridors that would keep Flight 25 separated at least sixty miles laterally from other jets crossing the Atlantic. On board were twenty crew members and 397 passengers, nearly a full load. Out of the corner of his right eye, the co-pilot caught a glint of something shining in the sunlight. As he turned to his right to look out the cockpit, he was shocked to see it was another airplane heading directly for their plane.

The co-pilot barely had time to call out a warning, and no time to take evasive action. Less than three seconds later, the plane, clearly another jetliner, flashed underneath the 747. The tail fin of the other plane may have come within thirty feet of striking the Continental jumbo. The two planes were so close that the plane passing underneath lost altitude as it encountered the turbulent bow wave of the slightly higher plane. Only a minute difference in the accuracy of the two altimeters (which measure altitude) aboard the two airplanes prevented a major disaster.

Radio communications quickly revealed the other plane to be Delta Flight 37, an L-1011 on its way from London to Cincinnati. The plane had been assigned to Track C, which should have kept it parallel to, but sixty miles away from, the Continental plane. Somehow, the plane had wandered off course by eighty miles.

Subsequent investigations found the Delta crew had misprogrammed their INS computers before they departed London. A single digit had likely been misentered into the computer. Proper cross-checking procedures that would have caught the error were not followed.

The Canadian Aviation Safety Board investigated the incident, since it occurred in airspace near Newfoundland. In reviewing the near-disaster it stated, ''The requirement to manually insert latitudes and longitudes on self-contained navigation systems makes such systems prone to input error.''

This problem is one of many that are due to the failure of airline training to keep up with the new demands of airliners that are largely software-controlled. The increased use of software demands new skills. Pilots also need to develop a strong awareness of how to handle software. The Board criticized both Delta and the pilots, the former for failing to emphasize the importance of cross-checking procedures, and the latter because they ''did not appear to be sufficiently aware of the importance of adhering to the required INS procedures.''

This incident points up one of the paradoxes of increased reliance on software-controlled computers in airplanes. As flight systems become more reliable, the anticipation of reliability may make pilots less able to spot errors. Pilots do what the software directs them to do, even if it's wrong.

The development of INS, for example, has greatly reduced, but has not eliminated navigation errors. The FAA has recorded numerous INS malfunctions or programming errors, at least one of which caused a jetliner to stray more than 250 miles off course. William Reynard of the FAA warned of pilots' growing reliance on the inherent accuracy of INS. Pilots were sometimes blithely following incorrect INS directions, merely because they expected the INS to always be correct. "This anticipation of correctness," Reynard said, "illustrates one of the paradoxes of modern technological development."

In the case of the Delta-Continental near-collision, the crew had ample opportunity to catch their error, but failed. As the Delta L-1011 cruised over the North Atlantic and crossed thirty degrees West longitude, the jetliner's autopilot began a shallow left turn. The course to Cincinnati took a gentle dogleg at this point, turning left ten degrees. But because of the misprogrammed INS, the plane turned twenty-six degrees to the left, putting it on a collision course with Continental Flight 25.

When the captain looked at his INS readout, it said it would be sixty-one minutes' flying time to the next checkpoint. According to calculations made before takeoff, this next leg was only scheduled to take forty minutes. Neither the three-man cockpit crew nor air traffic controllers noticed the difference. The crew never checked their heading against their pre-determined course.

In September 1989, a Varig Airlines 737 crashed into the Amazon jungle. Twelve passengers were killed. The cause may have been a single misplaced zero. Brazilian investigators concluded the pilot misentered the plane's intended course as 2700 instead of 0270. The Brazilian pilots' union disagreed and said the flight computer itself malfunctioned and issued faulty instructions. Either way, the two pilots incredibly flew for three hours before realizing the Boeing was headed south, instead of north. The plane ran out of fuel trying to return to Maraba, and was forced to ditch.

Crews are leaving it to the software to run things. And when it goes haywire they are not always equipped to take over. They are sometimes caught off guard, unaware of what their airplane is doing.

China Airlines Flight 006 began as a routine flight across the Pacific from Taipei to Los Angeles. Two hundred seventy-four people were aboard as the 747SP slowly made its way down the west coast of California. The plane was flying in the bright sunlight above the clouds at forty-one thousand feet, some three hundred miles northwest of San Francisco, battling some light turbulence.

The plane's number 4 engine, the outermost right-side engine, became ''hung''; although the engine was still operating, it was failing to produce much power. The giant 747 wanted to turn to the right, in the direction of the balky engine, in much the same manner as a car will tend to pull to the direction of a flat tire.

But the captain failed to notice the autopilot was compensating for this problem. When he shut off the autopilot, the plane swung violently to the right, and tumbled out of control. The giant plane went into a diving vertical roll. China Airlines Flight 006 plunged thirty thousand feet towards the Pacific Ocean for a harrowing two minutes before the pilot was able to regain control. The violent maneuvers ripped large chunks off the tail fins and landing gear, and permanently bent the wings. Miraculously, there were only two injuries and these were minor.

Veteran pilot and *Flying* magazine editor Peter Garrison notes the irony (vol. 113, October 1986). ''Except for 4/1000ths of an inch of excess wear in a fuel controller, there was NOTHING wrong with the airplane. Nor was there anything wrong with the crew; they were well-rested, well-trained, and highly experienced. China Airlines Flight 006 graphically shows what a misalliance the marriage of man and computer can sometimes turn out to be.''

Airliner software is often poorly designed in that it probably does its job TOO well. As Garrison notes, ''The crew's role is reduced to one of monitoring the performance of boringly reliable systems. Humans make bad monitors, however. To perform well, they need to be continually 'in the loop'; that is, they need to be giving commands or control inputs and then feeling or seeing the result.'' Too often, modern automated cockpit software is structured so that it gives the pilots too little to do when it comes to operating the airplane. When something goes wrong, the pilot suddenly is immersed in the midst of a bewildering situation.

Like many of his contemporaries, 747 pilot Harold Ewing worries that the increasing computerization of planes may be breeding dangerous complacency among pilots. As Ewing explains, ''The technology is beginning to create its own environment. You find really, really gross errors sometimes because people are relying to such an extent on the automated equipment. The pilot wasn't cross-checking the machine at all, because he'd convinced himself it hardly ever fails. What concerns me about the advent of this technology is that it is causing a dramatic increase in complacency, and an erosion of situational awareness. I'm concerned the next generation of pilots may not have that, because they will have grown up in an environment of such enhanced reliability. They might feel almost cheated or victimized if something goes wrong, because, 'gee, the equipment is so reliable it's not supposed to go wrong.'"

On February 28, 1984, a Scandinavian Airlines System DC-10 touched down at John F. Kennedy International Airport in New York. It was a rainy, foggy day in Queens. In the distance, the city's famed skyscrapers were cloaked in low, gray clouds that hung over their spires and sporadically paused to unleash bursts of chill rain and wind. Instead of braking to a stop, the plane skidded off the end of the runway and finally came to a halt when its nose burrowed into a marsh in Jamaica Bay. Controllers in JFK tower could not see the plane land because of the fog. But Ground Control's radar showed a blip that disappeared as the plane went off the end of the runway.

"Scandinavian 901, Kennedy. Are you okay?" he radioed. No reply. The controller quickly picked up a red phone connecting him with the airport fire station. Crash crews raced to the scene and found the jumbo jet resting like a forlorn whale, sitting in the shallow water and marshes of Thurston Basin, six hundred feet past the end of the runway. None of the 177 people arriving from Stockholm and Oslo were injured, though a few suffered exposure in the chilly waters, and many suffered a good fright as the plane slid off the runway, throwing up mud and debris.

SAS officials suggested the plane had hydroplaned off the wet runway. Yet there was no significant water buildup on the pavement, and the tarmac had tiny grooves cut into it to help braking. The DC-10 had made its final approach to the runway at unusually high speed. Pilots normally calculate a reference speed, or "bug" speed they hold to as the plane approaches the runway. The SAS plane crossed the runway threshold at a ground speed of 235 miles per hour, where normally it should only have been doing 180. The three-member cockpit crew noticed the approach was fast, but took no corrective action before the plane touched down. As the Canadian transport ministry noted afterwards, "The flight crew had no reason for concern because they had the autothrottle on. They had been taught that the autothrottle would take care of the speed of approach because it was, as its name implied, automated."

The fault lay in the software. The autothrottle had not been programmed to deal with wind shear conditions. A low-altitude shift in winds meant the engines should have been throttled back as the plane was on final approach. But the autothrottle, which knew nothing about wind shear, instead steadily increased the power. When the plane touched down, it was travelling far too fast.

Capt. Steve Last of British Airways explains. "The guys who designed the system had never contemplated flying the airplane under those circumstances [wind shear], which are in fact not that uncommon. It coped with the theoretical world that the person who envisaged it had in mind, but that, unfortunately is not what reality is."

Pilots must be constantly alert to new, more insidious software errors that are invading the cockpit.

British Airways, Northwest, and several other airlines noticed several incidents where the engine throttles on 747-400s mysteriously shut off by themselves in mid-flight. British Airways pilots noticed the problem on at least six flights, noting that it occurred several times on one flight alone. The problem was traced to a software error that mistakenly told the FADEC engine computers that the plane was on approach to landing. Boeing corrected the software error, but the problem persisted. This forced a second correction, which appears to have been effective.

The FAA discovered a potential problem with Honeywell Digital Air Data Computers on some McDonnell Douglas MD-80s. Some pilots were noticing that if they aborted a landing approach, and attempted to circle around for another approach, the computer sometimes fed erroneous information about the plane's speed to the autopilot. The error could have caused planes to lose too much speed and crash, although the automatic stall warning system in most cases would have alerted the pilot to the problem. The computers have since been modified.

British Airways has logged six instances where autopilots on Boeing 737s mysteriously changed the jetliner's altitude without warning. In one case a crew instructed the autopilot to climb to an altitude of eight thousand feet. The plane promptly levelled off at seven thousand feet. Puzzled air traffic controllers asked the crew what was going on. The computer autopilot had somehow changed its setting by itself.

The pilot reset the altitude hold mode for eight thousand feet and the plane dutifully climbed to eight thousand feet and levelled off. This time the pilots kept a watchful eye on the flight management system. As they were watching, the display for the target altitude again changed to seven thousand feet.

In every instance, the error was caught as flight crews checked the autopilot against other flight instruments. But the change in altitude was not always caught right away. The problem could easily wind up putting airplanes on collision courses.

Errors of this type are particularly dangerous because they may cause pilots to either act on incorrect information, or worse, start to disbelieve the software because they know it's prone to giving them false information. In the case of the wayward autopilots, the problem was traced to a faulty design in a computer chip aboard new 737-300s, and subsequently corrected.

As jetliners become more and more complex, airline pilots are ironically being taught less and less about how the airplanes work. Prior to the 1970s, pilots were trained nearly as engineers were. Training classes went into great detail about how the mechanical systems of airplanes operated. Pilots were taught how the systems could break down and how to cope under such circumstances.

But, as planes became more computerized in the 1970s, such detailed instruction became impractical. Planes were simply too complex. Training evolved into a system referred to as need-to-know. Instead of being taught the mechanical background of how a particular piece functions and how it could break, pilots are now often taught only what steps to take it if breaks down. The goal is to get a standard conditioned response from the cockpit crew to deal with emergencies.

Capt. David Walker, a flight instructor for Air Canada, refers to this type of training as, " "When the bell rings, I'd like you to press the green button and as a consequence of that a banana will drop out of the slot.' " He admits, "It's a little more complex than that we hope, but you had a specific behavioral objective which you wished the pilot to achieve and you taught him the steps to achieve that." Of less concern is teaching the mechanism that causes the failure.

Many pilots feel this need-to-know system leaves crews ill-equipped to deal with equipment failures, especially multiple failures or failures that are particularly unexpected. "Flight crews are only told as much about the airplane as they need to operate it," British Airways flight instructor Pat Farrell says. "But we are not being told enough about the airplane. When things malfunction we do not have the background knowledge to enable us to always make the correct decision."

Given the complexity of today's airliners, it is simply not possible for pilots to have in-depth working knowledge of all the plane's systems. And it's hardly reasonable to expect airline pilots to become software engineers. But a thorough working knowledge of the systems aboard the aircraft would leave flight crews better equipped to handle emergencies, and more importantly, to prevent them from occurring in the first place. In the incident where the Air Canada 767 ran out of fuel, Captain Pearson did not know that the fuel computer processor used two channels. If he had been taught that, he might have realized that he could have regained partial use of the fuel quantity indicators and known how much fuel he had by merely pulling one of the circuit breakers.

In the case of the China Airlines 747 that nearly tumbled out of the sky near San Francisco, the flight engineer misdiagnosed a minor engine problem. A situation that the Boeing flight manual does not even classify as an emergency nearly became a catastrophe because the crew member was apparently unaware of an otherwise unimportant engine part known as a bleed-air valve, a defect in which had caused the engine to lose power.

The solution is better training that teaches pilots how to deal with the complicated computer systems they oversee. Air France, for example, offers pilot training that provides more technical background into the airplane itself and how it operates. After a month or two of flying a new airplane, a pilot is taken back into the classroom where he takes additional courses to learn about the mechanical and electrical systems and how they function.

American Airlines also offers solid training that stresses teaching pilots how systems operate, not just what to do when they break down. American Airlines operations chief Robert Baker said in *Aviation Week & Space Technology* magazine (vol. 128, April 4, 1988), "We don't want the pilots to know that it's valve number XX. We want them to understand what the valve does, and if it doesn't do that, what happens."

It's interesting to note that American Airlines needed to change Airbus Industrie's training methods when it took delivery of its first Airbus A300s in 1988. American officials felt the Airbus Aeroformation system stressed memorizing technical data over developing a working knowledge of the airplane's systems. It took American a year to rewrite the Airbus Aeroformation training system to suit its own preferences.

Although increased computerization is designed to reduce pilots' workload, software needs to be developed that puts pilots "back into the loop" as an integral part of the system. Someone who does more than simply push the right button at the correct time. As Capt. Harold Ewing puts it, pilots acting by rote "will cause countless more accidents until we somehow find the collection of disciplines to get a handle on it. The reality is, the pilot's in the airplane to be the ultimate backup in case something does go wrong. And if they can grasp that and understand it and pattern their behavior accordingly, we can go through many technological revolutions without any problem. On the other hand, if they don't, we're going to have major, major problems. And I think our track record as a society does not support the view we're going to do a particularly good job.''

Editor J. Mac McClellan said in *Flying* magazine (vol. 117, May 1990), ''To fly safely in ever more crowded skies and to manage increasingly sophisticated airplanes with fewer people, we need automation. There are in fact some flying tasks that are too difficult for the human pilot to perform reliably every time, so we turn those tasks over to machines to help and protect the human. But the human pilot needs different training. We must learn to work with automated systems so that we're prepared to move ahead with technology. The modern airplane demands not only modern training, but modern attitudes towards automation. The human remains the primary backup.''

Caution must be the watchword. Engineers steadfastly maintain that increased use of software enhances safety by reducing pilot workloads and making systems more reliable. They will tell you that backups and system redundancies can prevent certain failures from ever happening. In truth, software can never be made completely reliable and trustworthy. But the problem is more than just the equipment and its software. As McClellan points out, ''Crashes in modern automated airplanes are happening with all systems, except the human pilots, functioning normally.'' Proper design of computers must not be ignored. Pilots and airlines must change their cockpit procedures and training, especially with a mind towards always expecting the unexpected.

Automation can be made to work safely, but only when carefully thought out. For as Capt. Harold Ewing offers, ''Anytime an engineer puts his arm around you and tells you not to worry because something can't possibly happen, then that's the time to be scared.''

Chapter Three

THE BIGGER THEY ARE...

THE world's most reliable software arguably belongs to AT&T. Software experts point to the company time and again as a shining example of how software can be written that performs complex tasks with clockwork-like dependability.

AT&T has been a leader in developing software that can operate around minor errors without suffering major, systemwide traumas. A good thing, too, because the nation's smooth flow of commerce and communications largely depends on the functioning of its long distance telephone service.

But even the most reliable of software contains errors. Predicting what effects they will have is as impossible as trying to predict when and where they will occur. And finding and fixing software errors can make searching for a needle in a haystack look like child's play.

Software errors can have surprisingly far-reaching effects. A single, tiny error managed to cripple a large part of the entire nation's commerce for nine hours.

IN a huge diamond-shaped building in Bedminster, New Jersey, a team of AT&T technicians stands watch over the nation's primary long distance phone lines. All told, the system is responsible for over a hundred million phone calls a day. AT&T handles more than seventy percent of the nation's long distance traffic. The nerve center of AT&T's Network Operations Center is a war-room-like setup where twenty-four hours a day, technicians man a bank of computer consoles. In front of them, a wall of video displays stretching twenty feet high constantly monitors the health and pulse of the long distance telephone lines that stretch across the country.

Bill Leach had to admit that for a Monday, this was a slow day today at NOC, or "knock" as employees called it. Today was a holiday, Martin Luther King, Jr.'s birthday, and so with banks and government offices closed, call volume was lighter than normal for a Monday. So far today, traffic was running about twenty percent below normal, typical for a holiday.

Anyway, Bill assured himself, the system pretty much ran itself. The video wall showed that, as usual, hundreds of thousands of calls were flowing smoothly along AT&T's electronic highways. The seventy-two video display screens winked on and off occasionally to show where the system was making minor automatic adjustments to keep the flow moving. The job of the network managers is mostly to watch and fine-tune the system—the operations center is the most highly automated phone switching center in the world.

On this day, Bill was one of about a dozen people assigned to various housekeeping tasks at NOC. The huge long distance system didn't need much tending to today. At the far left of the massive video wall, screens monitored the condition of AT&T's international network. Without banks or government offices being open, overseas traffic today was negligible. Dominating the center of the video wall was a bank of twenty-eight video screens, displaying the condition of the vast North American network. Most of the screens carried black cut-outs of the United States. As trunk lines became temporarily overloaded due to call surges or mechanical problems, a line would appear in blue, yellow, red, green, or magenta, depending on the severity of the condition. Computers then instantly would reroute calls to other lines that still had excess capacity.

To the right of the wall, screens showed the health of business services such as 800-numbers. Another bank of displays on the far right told the condition of AT&T's vast network of terrestrial microwave relay towers, coaxial cables, and satellite ground stations for relaying calls.

The video wall stretched an imposing two stories high, casting a faintly eerie glow across the cavernous NOC room and the twin rows of console operators in front of the wall.

The system was automated enough to deal with most problems on its own. Calls could be swiftly rerouted around natural disasters or errant backhoes, which lately seemed to be making a new sport out of digging up and severing AT&T's fiber optics lines.

Still, Bill had every reason to feel confident in the near-bullet-proof integrity of the system. The company, and NOC in particular, had in the past months passed two of its toughest tests ever with flying colors. Hurricane Hugo had struck Charleston, South Carolina with the full force of its 120-mile-an-hour winds in September. Fine antebellum Southern mansions that had withstood a hundred years of time were smashed into toothpicks, and sleek twelve-meter yachts were left stacked like so many toy boats. Barely a month later, as fans settled into Candlestick Park to watch the third game of the World Series between San Francisco and Oakland, an earthquake measuring 7.2 on the Richter scale struck the Bay Area of California. Although people soon became familiar with the televised sights of collapsed freeways and huge fires in the Marina District, phone service had remained remarkably intact. In both disasters, the software programming in the NOC computers had been able to keep long distance lines open by diverting call traffic. In the case of the earthquake, calls to and from California were rerouted through Denver.

Once again the phone system had proven itself the paradigm of reliability: the one thing people could count on. In addition, just a month earlier, the company, in its never-ending search for even greater efficiency, had completed a minor update to its software that did an even better job of rerouting calls in the event of a breakdown.

The long distance network made up one of the largest computer systems in the world. Actually, it was a complex series of computers. Scattered around the country were 114 switching centers that handled the routing of calls as they made their way across the country. Each switching center was a computer by itself; two computers really, one main and one back up, to ensure the system always ran smoothly, even when problems popped up from time to time.

The computer network did a great job of keeping an eye on itself. Every five minutes, the switching centers fed a flurry of data to the NOC, reporting on surges in calls and minor mechanical problems that could potentially disrupt call traffic. The information would appear on a display on the video wall, giving network managers a chance to make sure the system was taking care of itself properly.

Today the system was working to perfection. The switching system was smoothly handling the light volume of calls. On one of the video displays on the wall, the Cable News Network was on, as it always is. Bernard Shaw was saying something about new fighting in the Soviet republic of Azerbaijan. NOC technicians, although mildly interested in international politics, were far more interested in natural disasters that could hinder phone traffic, such as earthquakes or tornados.

But even Mother Nature was cooperating today, allowing the console operators to turn their attention to other matters. The Super Bowl was less than two weeks away. The flood of thousands of tourists into New Orleans, and the scores of journalists rushing to file their stories at the same time would mean added demands for AT&T's services. The Chinese New Year was coming up, which would mean more traffic to the Far East. Up on one of the video screens was a calendar listing significant upcoming events. AT&T liked to plan ahead for any potential strains on the system. NOC hates to be surprised.

There had been few surprises today. Bill Leach was sitting comfortably at his desk overlooking the console area. Through the large glass window pane in his office, the view was of network managers coolly going about their tasks. Bill, for that matter, had his nose buried in a stack of articles he'd been wanting to catch up on. With the holiday today, things were nice and quiet. A good time to catch up on some technical reading. In a few hours, Bill thought, he'd be on his way home, and he wouldn't even have to fight the usual traffic.

Out of habit, Bill glanced up at the video wall. It was 2:25 P.M. Every five minutes, the computer displays refreshed themselves and showed updated information on any trouble in the network. The last time he had looked, the board was quiet. The outline map of the United States in the center of the wall had been dark.

Oh-oh. Not anymore. Looks like trouble in New York.

One of the video screens showed a minor problem with one of the New York City Switching Centers, one of the 114 regional centers that handles the routing of long distance calls. A dot had appeared on the map, and it radiated red-and-blue lines. For some unknown reason, the circuits were full, and calls were being rejected.

"It wasn't that spectacular or alarming for that matter," Bill recalls. Minor disturbances in the system were an unwelcome but common nuisance. Switching centers would often back up as minor mechanical problems or a surge in calls created a temporary overload.

Bill looked at the network managers seated in the console area. *They ought to be able to take care of this without too much trouble.* The problem would be transparent to AT&T's customers. The software on the signaling network would instantly detour calls around the trouble area. No calls would be lost, and people making phone calls would never even notice there had been a problem. Anyway, if the trouble turned out to be more serious, the managers would simply reroute calls around the trouble area, an action which was second nature to them. Such problems seldom lasted more than a few seconds. Bill returned to his reading.

After another five minutes, Bill looked up again as the screens refreshed themselves. He expected to see the New York trouble indicator disappear from the screen.

This time he knew there was real trouble. New York City was

still on the map. But as the new display came up, New York was now joined by three other switching centers showing trouble. Atlanta. St. Louis. Detroit. All displayed similar red-and-blue lines, showing full circuits. "Now that was very alarming," Bill said later. "Because that just doesn't happen."

But it was happening. As one technician put it, "We knew right away we were in trouble, because we were dealing with something we'd never seen before." Bill swiftly moved from his office into the heart of the Network Operations Center. Console operators were as puzzled as he was. They immediately began rerouting calls around the troubled areas. But instinct told Bill Leach this would only be treating the symptoms of what ominously threatened to be a much larger problem.

I don't know what's going on here, but we're going to need some help. Fast.

Bill picked up a telephone receiver at one of the consoles, and quickly patched a call through to NESAC, the Network Switching Assistance Center out in Lyle, Illinois. Technicians there provided technical support when there were problems that couldn't be handled at NOC. This one definitely qualified. Leach swiftly outlined the problem, but NESAC technicians were equally at a loss to explain what was happening.

Bill kept a watchful eye on the video wall as he described the rapidly deteriorating situation. As he watched in horror, things got worse; the fault cascaded through the entire AT&T network. Fault indicators spread across the map like a runaway cancer.

"Things were going downhill on roller skates. Something extremely unusual was happening. Each succeeding five minutes, it just got worse. The lines continued to spread across the map."

Bill grabbed the telephone receiver on his console and stabbed a code on the touchpad that connected him with James Nelson, the Network Operations Center's chief. "You better come down here," he said. "I think we've got a big one."

Calls were being turned away by the millions, even though AT&T's phone lines were only half-full. Mechanically, the system was fine. But some invisible gremlin was tying up the lines. Within a matter of minutes, the entire nation was in telephone gridlock. The experience would be new for everyone: the country had never before experienced a telephone problem on this scale. The impact on businesses was both instantaneous, and for many, catastrophic.

In Omaha, Nebraska the lunch hour was just winding down. It was clear and warm outside. The city was enjoying unusually balmy weather this January, and the downtown sidewalks were free of snow. Chuck Waymire was at work in his office at Mid Trans-America, a company that handles trucking permits. The first indication something was amiss came when workers came into his office and told him the phones had stopped. Puzzled, Chuck asked what they meant. They told him phone calls had suddenly stopped coming in, and they could not make calls outside the local area.

The long distance telephone is Mid Trans-America's lifeline. Twenty-two thousand trucking companies around the United States and Canada count on Mid Trans-America to obtain temporary truck permits for their rigs, and then fax the permits directly to them via long distance phone lines. Delays mean rerouted shipments and lost money.

Chuck tried to make a long distance call and got the familiar "all circuits are busy, please try again later" recording. He quickly called local telephone operators in Omaha. They relayed the bad news: because of technical problems, long distance calls were not going through. Chuck hung up the phone and turned to his workers. "We're completely shut down," he said. "We're dead in the water."

Mid Trans-America would not complete another long distance call that day. A similar scene was being played out in millions of businesses from coast to coast. In twenty-two years of operation, Mid Trans-America, like many other businesses, had used only one long distance company: AT&T.

In Tulsa, Oklahoma, home to the nation's largest travel reservation computer, activity inside the sprawling American Airlines Reservation Center slowed to a crawl. Its massive Sabre computer system typically handles more than one million reservations from business travellers, tourists, and travel agents every day.

But suddenly, the volume of incoming calls dropped sharply. American estimates by the time the day ended it had lost two-thirds of the calls that customers tried to make to its reservations center. The net loss: an estimated two hundred thousand phone calls.

Other travel companies quickly joined American in learning a painful lesson in how vulnerable their 800-service was. United Airlines, which operates the second-largest reservations system, estimates thirty percent of its calls were hung up. Idle reservation agents sat at their terminals and reconfirmed reservations for sold-out flights for the upcoming Super Bowl in New Orleans. Others used outgoing lines, some of which were unaffected, to call out to advise passengers of schedule changes.

For the nation's telemarketers, in a nation swept up in the craze and convenience of shopping by telephone, and even by television, the crash of the telephone system was a blow to the jugular. Steve Idelman, chairman of Idelman Telemarketing, watched as the vast majority of his phone lines turned from revenue-producing links to the outside world, to useless copper wires. One of the nation's largest marketers of products by telephone had been cut loose and set adrift. Idelman sent eight hundred of his workers home. The loss of long distance phone service for just one afternoon cost his company an estimated seventy-five thousand dollars in lost revenue.

At Dial America, branch manager David Haller wasn't completely dead in the water, but he was definitely treading water to keep his business afloat. As he struggled to determine the exact nature of the problem, he was grateful that years earlier he had decided to use two long distance services. At first, he thought the problem was affecting all long distance companies. But as his telemarketers attempted more calls, it became clear only his AT&T phone lines were affected. Still, it meant he had just lost thirty percent of his capacity. None of the calls on AT&T lines were being completed.

Haller decided he had no choice but to make do with what he had. "If we had been one hundred percent AT&T," he said, "we would have been devastated." As it stood, he called it "the most devastating failure I've ever seen in telemarketing." Haller sent a third of his workforce home early. The phone tie-up, he estimates, cost his company perhaps eleven thousand dollars in revenue. "Thirty percent of our phones were out of service. We couldn't get any long distance out on them."

At the New York Stock Exchange, which relies heavily on AT&T to handle its enormous volume of transactions, there was little interruption as traders on the floor went about their business. Stock exchange officials, who had wisely recognized their vulnerability to a breakdown in the phone system, executed carefully laid plans, and simply switched their phone and computer data lines over to an alternate long distance service.

Jim Nelson finally arrived at the heart of the AT&T Network Operations Center in New Jersey. He had hurried downstairs after receiving Bill Leach's urgent call that there was trouble. He was stunned by what he saw as he entered the room: the entire map was lit up like a Christmas tree, indicating trouble along the entire network.

The network was still operating, but the situation was chaotic. Switching centers around the country were turning themselves off-and-on at random. Trouble indicators lit up across the entire screen. Many of the 114 switching centers around the country radiated red-and-blue lines, creating a starburst effect. Every five minutes the status maps changed as the video screens updated their data. The indicators disappeared, only to be replaced by an entirely new set of starbursts, showing trouble at different locations. The screen resembled a giant pinball machine gone crazy.

Bill Leach gave Nelson the bad news: console operators had already tried using a number of standard procedures to revive the network. They had always worked in the past. This time they had no effect.

The network had not completely shut down, but more than half of all calls being attempted by AT&T customers were now being rejected, as the recalcitrant computers refused to allow them entry into the phone lines. The situation was not staying in one place long enough for technicians to get a clear picture of what was happening. "It was a dynamic situation," one AT&T worker said, "a switching center would be up one minute and then down the next. Nothing stood still long enough for us to pinpoint what was going on." Console operators stared blankly at their computer screens, stunned by the speed and sheer magnitude of the collapse.

Nelson and Leach moved away from the chaotic video wall to a bank of computer terminals that could provide more detailed information on what was happening. The terminals could explain the status of each of the 114 switching centers around the United States. The screen would show a designator number of each center, and next to that, error messages that were tumbling out of each switch. The two men examined the avalanche of error messages that was coming up on the computer terminals. Somewhere in there, they knew, lay a pattern that would tell them what was troubling the system. But deciphering the pattern was beyond them. For this, they would need specialized help. Nelson put out an immediate long distance conference call to scientists at the world-renowned Bell Laboratories. AT&T's private long distance lines were, fortunately, still operating normally. Bell Labs, Nelson knew, had some of the finest engineers and computer scientists in the world, and if AT&T ever needed them, now was the time.

The atmosphere inside NOC was strangely subdued and quiet. The scores of red-and-blue trouble indicators on the video wall added to the soft fluorescent glow inside the center. Console operators were busily trying to reroute as many calls as they could. People realized shouting at the computers was not going to solve the crisis. They had been trained to handle crises like this. But people were nearly panicking beneath their calm demeanor. The supposedly unsinkable *Titanic* of telecommunications was going down.

Unknown to anyone at the time, the problem had started inside one of the phone switching systems in lower Manhattan in New

York City. Just before 2:25 that afternoon, a minor mechanical malfunction tripped the software into reset mode. The problem was a trivial one, and a common occurrence. Such malfunctions occur often, but are hardly ever noticed by the public, as the system is designed to smoothly handle such problems without any disruption of service.

The switch went into what's known as a fault recovery mode. The 4ESS unit temporarily took itself out of service until it could straighten itself out. Its built-in maintenance software began correcting the problem. During that time, no new calls could be handled.

The switch sent out a message to all other switches around the country, alerting them that no new calls should be sent to New York during this interval.

Six seconds later, the New York switch completed its reset. The switch now began sending out new long distance calls. As other switching centers around the country from Philadelphia to Los Angeles received the new calls, they took note and began updating their software programming so that they could begin rerouting calls back to New York. While this updating was taking place, a series of new long distance calls arrived from New York. The calls were spaced in a particular pattern, less than 1/100th of a second apart, that somehow triggered a fault in the software, disrupting some data. The software switched over to backup computers, but they quickly fell victim to the same problem. At switching centers around the country, the problem was now beginning to snowball. Computers began taking themselves out of service. As they came back on line, they sent out new long distance calls that in turn knocked out their neighboring switches. Soon, "I'm out of service" and "I'm back" messages from all 114 switching computers around the country were bouncing back and forth around the network. The huge network had turned into a schizophrenic circus of computers shutting themselves off-and-on.

It was like watching a hundred mud wrestlers crowded into a too-small arena. As each player falls and then gets back up, he knocks over people standing next to him. As those players rise to their feet, they in turn topple a new set of people who themselves have just gotten up. The switching computers were shutting each other down. As each computer came back on line, it was sending out messages that knocked out neighboring switches, and the cycle continuously repeated itself around the nation.

The network was now in a mode of random chaos. It was impossible from one moment to the next to predict which switching centers were working and which were shut down—the situation was constantly changing.

Nelson and Leach began to search like detectives for clues. The video wall showed that some of the trunk lines that carried calls between cities were experiencing problems, but others were not. The video maps only monitored the condition of the entire network. Bill could use a computer terminal to call up ''machine-discretes''— information about specific switching centers or trunk lines. He carefully began analyzing which lines seemed to be free of trouble.

AT&T had been in the process of converting its massive call switching network. An old system called Signaling System 6 in use since 1976 was being replaced by a new high-speed computer switching network known as Signaling System 7.

Bill tapped his keyboard and called up information about a trunk line that was still working. It used Signaling System 6. And another one. The same thing. A third line. SS6. Bill reversed the procedure and found all the troublesome lines were using Signaling System 7.

Bingo.

''We quickly discovered it appeared to be a System 7 related problem,'' he said later. ''Any trunking that used that type of signaling seemed to be affected, while the older version was not.''

Ironically, all the thousands of AT&T computers, trunk lines, and switching relays were in perfect operating condition. The network had been crippled by something far more insidious than a mechanical failure: a tiny flaw in the invisible operating instructions for the computers that ran the switching centers. Buried somewhere deep within the millions of lines of numbered codes that ran the switching computers, lay an error that had somehow gone undetected for months, and was making the network run amuck. Finding it would be a Herculean task. It would be, as one technician described it to the *Wall Street Journal* (vol. 215, February 14, 1990), like "looking through a week of *Wall Street Journals* to find a missing dot above an *i*."

The engineers had no choice. They set about their immense task. Network Operations Center Chief Jim Nelson, Bell Labs Vice President Karl Martersteck, and William Carroll, Network Operations Vice President, quickly huddled and drew up a game plan. Bell Labs maintained research facilities in Illinois, Ohio, and New Jersey. Each facility had played a different role in developing the network. In Naperville, Illinois, where the 4ESS switch had been developed, scientists guessed correctly the problem was somewhere in the software. In the 4ESS System Lab on the third floor of the facility, they maintained working models of the switching network. Normally, the switch in the lab is used to develop and test new switching software. All such work was hastily put aside. The faulty software was quickly loaded into the 4ESS laboratory system in hopes technicians could narrow down the problem. As Bell Labs spokesman Karl Blesh related, "We knew the problem was somewhere in the switching system, but we didn't know exactly where." Together with scientists in Columbus, Ohio who had developed the components of the system, and in Paramus, New Jersey where the system had been designed, the Naperville team set out to try and recreate the disaster in a controlled setting.

Network Operations Center manager Bill Leach looked at his watch. It was barely three o'clock. It was going to be a long day.

ACROSS the country, AT&T customers were scrambling to find alternate phone lines. Rival companies such as MCI and U.S. Sprint maintain a separate network of lines that were unaffected by the troubles at AT&T. But customers seeking refuge quickly ran into a roadblock ironically set up by AT&T itself. AT&T operators are under standing marching orders not to give out the touch-tone access codes that allow customers to patch into rival services. Customers were turned away in frustration. Inexplicably, AT&T officials did not reverse the order until three and a half hours into the crisis. By that time, many east coast businesses had already closed for the day.

For businesses in western time zones, the struggle continued. At the Azer Travel Agency in Los Angeles, Oriette Mani found it took dozens of tries to place calls to make reservations.

Some AT&T customers were less hard hit than others. J. C. Penney operates sixteen order-processing centers, where a legion of seven thousand operators handle catalog orders from around the country. Calls continued to flow in at near-normal rates. Duncan Muir, a J. C. Penney spokesman said, "Virtually all of our telephone business is done through AT&T. But we really didn't have a problem. Our contract service with AT&T was virtually unaffected."

But other AT&T customers were not so lucky. Hotel chains, car rental agencies, and ticket centers lost thousands of reservations as frustrated customers tried and then tried again in vain to reach reservations centers. Holiday Inns watched as incoming calls to its centers in Utah, Illinois, and North Carolina dropped by ten percent. At the Physicians Mutual Insurance Company policy owners' services office in Nebraska, customer service representatives stared at their phones as call volume fell by a stunning seventy-five percent.

Other businesses managed to dodge the bullet completely. Several large retailers, such as L. L. Bean in Freeport, Maine, had previously switched over to competing long distance services, attracted by the lower price offered by companies such as MCI and U.S. Sprint. They were now grateful they had made the switch. In Minneapolis, one of the nation's largest mail-order liquidation companies, Damark International, heard about the AT&T problem through news broadcasts. The company's flow of seventeen thousand calls a day continued uninterrupted. Damark had switched to

Sprint two years earlier. Director of Teleservices Karen Beghly greatly understated the situation when she said, "We couldn't be happier with their service." Her team of two hundred operators was still taking orders for lamps and exercise machines as fast as they could handle them, while competitors cursed at banks of phones that were now useless.

The disruption of phone service across the nation was big news. But, ironically, producers at CBS News in New York found they couldn't get through to their bureaus scattered across the country to cover the one big story of the day: the crippling of the AT&T system.

As late afternoon faded into early evening, AT&T technicians at the NOC in New Jersey were able to hold their own. Since the older Signaling System 6 portion of the network seemed to be unaffected, as many calls as possible were rerouted to those lines. It was not a complete solution. Only thirty percent of the nationwide network still used the old SS6 system. Many cities had been completely converted to the newer SS7. A limited number of calls could be carried, although with some delays, but at least some traffic was flowing. But most 800-number businesses, such as airline reservations, car rentals, hotel chains, and mail-order businesses had already been converted to the new system.

Technicians from Bell Labs had been taking a close look at the messages the switching computers were sending to each other. The answer lay somewhere within the high-speed electronic jabbering the computers spoke amongst themselves. Precisely why the 114 switching computers around the country were shutting each other down remained a mystery.

Around 6:00 P.M. Bell Labs called Network Operations in New Jersey. The technicians still had not found the cause of the problem. But they were ready to try a fix that, hopefully, would restore the network to normal operating conditions. They had concocted an overwrite, a software "patch," a new set of instructions that would hopefully fix the system. The patch was loaded into a switching center computer, in hopes that if it didn't solve the problem, it would at least help pinpoint the cause of the malady.

The job of the NOC technicians was to observe: their role would now be to monitor the system and report back to Bell Labs if the patch seemed to be working. NOC network managers held their breath. The software patch failed.

FOR AT&T, which had built its reputation on an heretofore untarnished record of quality and reliability, the disruption could not have come at a worse time. Once the undisputed king of long distance telephone service, AT&T had seen its market share slowly but steadily eroded by new competition. Newcomers MCI and U.S. Sprint, in particular, took advantage of deregulation and new digital technology to mount a serious challenge to what remained of Ma Bell. Although still number one, AT&T had seen its share of the long distance market dwindle from a dominant ninety percent in 1985, to seventy percent at the time of the crash.

AT&T had been aggressively fighting back with a multimillion dollar advertising campaign that stressed the reliability of its service. AT&T chose not to compete dollar-for-dollar in price with its younger, leaner competition. Such a business strategy could evolve into a price war and erode profits. Besides, AT&T officials reasoned, they could offer something the competition could not: an untarnished history of reliability.

But all that was about to change forever. On this particular day, AT&T proved to be anything but the right choice. In the wake of the disaster, AT&T would be forced to change its advertising strategy—to one that emphasizes price competitiveness.

Bell Labs engineers had traced the problem to a software error in a computer switching system that finds the most efficient way to route calls. As is often the case with software problems, the bug was in a new program that was designed to be a significant improvement over the old network.

In 1988, AT&T set out to improve its handling of calls. A new computer system called Signaling System 7 was installed that would connect calls faster, and handle twice as many calls as the previous system. Pressured by increasing competition in the new deregulated environment, AT&T sought to hang on to its dominance in long distance.

In 1976, AT&T engineers pioneered a new type of switching technology called out-of-band signaling, or known in the industry as common channel signaling. When you make a call, important information *about* the call, such as the number you're calling and where the call will be billed to, is separated from the actual transmission of the call itself. In this way, the information needed to route your call can be sent on ahead of the call itself, in much the same manner as an army battalion can send a speedy scout ahead to search out paths.

When a long distance number is dialed, the system looks at the number you've dialed and searches the network within thousandths of a second to see if a path is available, or if the party at the other end is busy. Previously, the entire call had to be routed in order to determine if the call could be completed.

In addition, the new system reduced the time delay between when you finish dialing and the first ring. The thirteen to twenty second lag was cut down to between four and six seconds. Although a savings of perhaps ten seconds might seem trivial, when multiplied by the hundreds of millions of calls carried each day, it amounted to a tremendous gain in capacity for AT&T lines. Out-of-band signaling also speeds credit card calls by allowing instant computer verification of calling card numbers. Verification formerly had to be performed by operators. It also allows new 800-number services that can route calls to different service centers on a moment's notice, and a host of other services that are still being developed. Telecommunication systems around the world, and even AT&T's primary rivals in the United States, all use some form of out-of-band, or common channel signaling.

But common channel signaling relies heavily on its highly complicated software. The fault lay not in the switching computers themselves, but in the software. And AT&T had inadvertently introduced the error into the system in an effort to improve it.

In the middle of December 1989, AT&T made a seemingly minor change to its switching software. The network had been plagued by a minor but persistent bug: a few calls were occasionally being lost when automated call switching centers came back on-line following minor mechanical malfunctions. The software program was updated, and worked flawlessly for the first few weeks after installation. Then, software engineers say, some particular combination of calls unforeseen by programmers triggered the massive shutdown. As Bell Labs spokesman Karl Blesh said, ''The probability of it happening was very low. The combination of events that happened to cause this was extremely unlikely.''

Further, the system had been specifically designed to avoid having a single failure shut down the entire system. Ironically, AT&T's computer division had been pioneers in designing a particular type of computer architecture that prevents computer malfunctions from causing runaway network failures. Supposedly, at least. Nearly every piece of equipment is duplicated by an identical backup system. This type of fault-tolerant system architecture is designed to automatically switch to backups to go around faults. The duality usually ensures continuous service, even through breakdowns. But it was vulnerable to a software error because it struck both the primary switching computers and their backups simultaneously. Afterwards, AT&T chairman Robert Allen noted with great irony, ''This was a condition that spread throughout our network because of our own redundancy.'' By safeguarding the system against any single failure, the company had failed to anticipate that a software error could strike all the switching computers at once.

Although no one could have predicted the failure, it was clearly a foreseeable crisis.

Late in 1989, the National Academy of Sciences released a report titled ''Growing Vulnerability of the Public Switched Networks.'' The panel noted that while the nation was becoming more and more of an information society that relied on telecommunications, the telecommunications networks "are becoming more vulnerable to serious interruptions of service."

John McDonald, executive vice president for technology of Contel Corporation, and chairman of the NAS panel, warned ominously in the *New York Times* (vol. 138, May 25, 1988), "The network is getting thinner and thinner and switches are getting bigger and bigger. When a major outage occurs," he warned, "it's going to cover bigger and bigger areas."

An AT&T official argued, "I think that perception is wrong," claiming the system had become more reliable, not less. Joseph Nacchio, director of AT&T's Network Engineering and Operations defended the system, telling the *Times,* "We have more route miles, more circuit miles, and more alternate path capabilities than we had before." Nacchio said the new equipment would be more reliable than the older technology it replaced.

The report highlighted one particular weak point in the system, the growth of common channel signaling, which relies on software of "enormous complexity" that increases the vulnerability of the system to disruptions. The panel was concerned because "the trend toward common channel signaling is irreversible and [therefore] network vulnerability is increasing. Without signaling, networks cannot function." Common channel signaling would prove to be the source of the AT&T crash.

Few heeded the warnings of the NAS panel, despite its further warnings about the vulnerability of the nation's telephone system to computer hackers and terrorism. People had relied on the telephone system for years, and continued to hold up AT&T as a shining example of dependable, never-fail technology.

A recent report of the House Committee on Science, Space, and Technology blasted the reliability of software programs in the military, but pointed to one example of how to do things right. The report concluded, "Keeping the nation's telephone system operating is a difficult and demanding task. It is not surprising that AT&T has produced pioneering work in software reliability." In hearings on the complexity of the Star Wars anti-missile defense plan, Pentagon planners had cited AT&T's long distance switching as an example of complicated software that works reliably.

But, ironically, the crisis of January 15, 1990 was not even AT&T's first nightmare in dealing with painful software bugs. Earlier problems had been discovered in order-processing computers. In October 1984, some 87,000 businesses around the country were startled to learn that they had been switched from their long distance carriers to AT&T. The switch affected some 285,000 phone lines. Perhaps the biggest indignity was heaped upon AT&T rival U.S. Telecom. The company had selected its own service, quite naturally, for long distance service at its headquarters in Kansas City. But the company suddenly discovered it had been switched over to AT&T. The Justice Department was called in to investigate. It concluded that AT&T had "failed to exercise sufficient care" in processing orders and that its computers had assigned AT&T lines to the wrong customers. AT&T was forced to shut down its computerized order processing for two months.

AT&T placed the system back in service, only to promptly watch it go wacky again. Some 117,339 letters were mailed out thanking people for choosing AT&T as their long distance carrier. Calls from angry consumers lit up AT&T phone lines. The letters had been sent to customers belonging to other long distance carriers. The problem was once again traced to a computer error.

Company officials defended the mistake, saying such hang-ups were to be expected given the tremendous volume of orders AT&T was handling. "Our error rate is less than one percent," claimed one AT&T official to *Business Week* (no. 2927, December 30, 1985).

An executive at rival MCI complained in turn, "Maybe 117,000 mistakes is within AT&T's margin of error, but to us, that's a lot of customers." The executive estimated the errors cost MCI more than $10 million in lost revenue. "How many computer errors are they going to get away with?" he asked.

* * *

AT 9:00 P.M., Bill Leach finally got the call he had been waiting for. The hundreds of Bell Labs scientists assigned to the problem had a new software patch they thought would work. Several different combinations had been tried in the preceding hours. Although they had not solved the problem, the scientists had used a process of elimination to narrow their search for answers.

The overwrite was hastily loaded into a switching center in California. Restoring full service now would still do some good for businesses in the Pacific Time Zone. Officials at the Network Operations Center again held their breath. Upon examining the data coming from that switch, NOC network managers declared that the overwrite appeared to be working.

The hastily devised fix did the job. The software patch prevented the switching center from sending out the "I'm back" message that was knocking out other switching centers. The overwrite was quickly loaded into all 114 switching centers around the United States. Within minutes, the red-and-blue trouble indicator lines began very slowly disappearing from NOC's electronic maps as magically as they had first appeared. Bill Leach was optimistic that a solution had at last been found. But he was nervously cautious at best. The entire episode had been so unpredictable; he would hold back from saying it was over.

Every five minutes as the video wall updated, trouble indicators vanished. It was a slow, tedious process as all 114 switching centers came back to a healthy condition. Finally at 11.30 P.M., some nine hours after the worst nightmare in AT&T's history began, the final starburst winked off the screen. The network was operating normally again. No one at the Network Operations Center cheered or jumped up and down. Bill Leach heaved a sigh of relief. It was time to go home.

JANUARY 16, 1990, dawned with the nation's phone lines up to full speed. The exact damage from the nine hours Ma Bell went temporarily insane will probably never be known. Hundreds of millions of dollars were likely lost as businesses failed to make connections with clients and customers. AT&T alone estimates it lost $75 million in revenue simply from lost calls. The long distance system had never stopped functioning completely, but of the 138 million long distance and 800-number calls placed that day, 70 million did not go through.

AT&T chairman Robert Allen said, "Even though it was a one-time hit to the network, it was certainly the most far-reaching service problem we've ever experienced. We didn't live up to our own standards of quality. We didn't live up to our customers' standards of quality. It's as simple as that. That's not acceptable to us."

Early in the morning, a group of top AT&T officials met to discuss strategy. The technical crisis was over; the public relations crisis was not. They agreed AT&T would have to take full responsibility for the crash. Attempting to hide or minimize the problem would only encourage wild speculation and public distrust. AT&T would have to take its lumps, but probably come out the better for it in the long run.

Some type of a conciliatory gesture was also needed to make up for the inconvenience the crisis had caused the public. It was agreed that for the first time in its history, AT&T would offer a full weekday of reduced rates. The date chosen: Valentine's Day.

When all concerned parties within the company had been briefed on how the company would handle the crisis, Chairman Allen stepped in front of the cameras to address the assembled media and explain why the nation's phone system had crashed. Although clearly embarrassed and apologetic, Allen made it clear in his post-mortem on the crash that AT&T was not liable for any losses to businesses resulting from service disruptions. Those having 800-numbers would be paid between twenty and fifty dollars, as specified in their contracts, for each toll-free line that had gone out of service. AT&T had guaranteed its customers that they would be switched to working 800-lines within one hour of any disruption of service.

*   *   *

AMONG AT&T's competitors, U.S. Sprint was quick to gleefully take advantage of the company's misfortune. That morning, Sprint rushed into print full-page newspaper ads admonishing businesses that got caught in the telephone snarl. "An important message to everyone whose telephone is the lifeline of their business," the ad warned, "Always have two lifelines."

For Sprint, which had suffered through an AT&T commercial mocking Sprint's "so clear you can hear a pin drop" ads by showing a pin dropping in the midst of a bank of operators idled by a loss of service, the ad was sweet revenge. What AT&T had warned Sprint customers to beware of, an interruption in service, had turned around and bitten AT&T itself.

U.S. Sprint should not have been so quick to gloat. The company had itself been a victim of some persistent software bugs. For the first few years after it began operations in 1986, Sprint was plagued by a costly series of computer billing errors.

In March and April 1986, defective new software was loaded into ten of Sprint's fifty-eight switching centers. The result: calls regularly went unbilled and millions of phone calls were given free rides on Sprint's lines. It's estimated the error cost Sprint between $10 million and $20 million. The bug was found and corrected.

The very next day, the software began a new series of misbillings, because programmers had failed to take into account the switchover to daylight savings time in calculating the start of evening discount rates.

In 1987, one clever consumer discovered he could bill a long distance call by dialing any fourteen-digit code at random. Who ended up being billed for those calls remains a mystery.

By July 1988, the problems at U.S. Sprint were so widespread that the Federal Communications Commission threatened to levy fines. It was concerned by the number of complaints it was receiving and it questioned the adequacy and accuracy of Sprint's responses to these complaints.

MCI, the nation's second-largest long distance carrier, chose to take the moral high ground, and refused to trumpet AT&T's failure.

And rightly so. MCI had once narrowly escaped a similar problem. Lightning struck one of MCI's computerized switching centers in Fort Lauderdale, Florida. The strike jammed five signal transferring computers and disabled a key network communication switch. Fortunately for MCI, however, the system was able to successfully reroute calls and maintain service.

AT&T officials acknowledged that the problem had been their responsibility. They patiently explained precisely what had happened, and admitted that it could happen again. As an AT&T spokesman explained, "Obviously there will be things we'll learn from this. We hope our honesty and forthrightness is preserving the confidence of our customers."

Customers were, indeed, mostly forgiving of the aggravation and lost business AT&T had caused them. Steve Idelman of Idelman Telemarketing shrugged and said he wasn't angry at AT&T, merely grateful that things were running again and that he was no longer losing money.

The incident apparently cost AT&T little in goodwill among its most loyal customers. Chuck Waymire of Mid Trans-America in Omaha reflected, "I'm very loyal to AT&T. I've always used them. Once in twenty-two years? That's not too bad. I'd like to have as good a record."

Outright defections were few. AT&T, by most estimates of Wall Street analysts, lost perhaps a bare one or two percentage points of its business to its rivals. But in the high volume world of long distance telephone service, that could still amount to as much as a whopping $1 billion in lost business annually, as shaken customers either abandoned AT&T or else signed up with other long distance services as a backup against future problems.

The phone crash had been a stunning blow to AT&T's previously spotless record, not only because of its breadth, but also because of its unexpectedness. Few had ever contemplated what would happen if the nation's phone system went down.

AT&T engineers removed the faulty software from all 114 switching computers and replaced it with the old program. Engineers at Bell Labs began their line-by-line, code-by-code examination of the program to determine how the fault had escaped their notice. Through tedious experimentation over the next several days, they were finally able to reproduce the chain of events that led to the January 15 crash. The engineers isolated the flaw in the program and corrected it. The new program was vigorously tested, as the previous version had been, to ensure it worked properly. The newly revised software was placed back into the AT&T network a week later, and has worked flawlessly since.

AT&T engineers are continually looking at ways to further improve the program. They hope they can further cut both the time it takes to complete a call and the chances of an error. But they insist there is no way they could have foreseen the crash. Says spokesman Karl Blesh, "It's impossible to test for one hundred percent of all possible conditions. There are so many conditions that you just can't test for every conceivable situation."

In short, AT&T was saying it could happen again. Only next time, it could be MCI. Or U.S. Sprint. Or some large corporation's private telecommunications system. Virtually every major telecommunications network in the world uses some version of the Signaling System 7 software to control its call switching.

To operate, Signaling System 7 requires an enormously complicated piece of software. A typical business word processing program might encompass thirty thousand lines of computing codes, but a typical SS7 configuration contains up to ten million lines of code. By the year 2000, it is forecast that "megaswitches" might require up to one hundred million lines of code.

The software is growing in complexity beyond the ability of programmers to properly test it in all the conditions it will have to encounter once it begins operating. The reliability of these systems is quickly becoming an unknown quantity. As the National Academy of Sciences noted, future call switching programs will be "sufficiently difficult so that confirmation of software performance in all networks modes and conditions may prove unattainable.'' Ironically, a top engineer from Bell Labs served on that NAS panel.

Such problems are not limited solely to long distance networks. Computer software that controls local phone calls has also proven to be been prone to bugs. In 1987, three of the regional ''Baby Bell'' phone companies were struck by software errors.

On September 11, nearly all of downtown Minneapolis turned into one giant busy signal. At 4:00 A.M., Northwestern Bell technicians had made some routine changes to software in a switching computer known as the IA Attached Processor. The computer refused to restart with the new software in place. Even reinstalling the old software failed to solve the problem.

Fifty thousand telephones, nearly half of all phones in the sprawling downtown, were out of service for four and a half hours. Hotel reservation clerks and car rental agents sat around doing nothing. The Minnesota Twins ticket office found its calls dropped off to a trickle of 192, instead of the normal 1,500 for the day. Only messenger services reported brisk business as businesses scrambled to have messages delivered by hand. Even when service was finally restored, it was plagued with delays in connecting calls.

On June 29, calls were backed up in nearly 1.1 million homes throughout Imperial County in California, which includes San Diego. A software bug in a switching computer tied up phone calls throughout most of the 619 area code. The problem lasted most of the morning and afternoon.

A similar problem struck Poughkeepsie, New York on July 18, knocking out or limiting service to fifty thousand customers.

EVEN the most optimistic of Bell Labs engineers cannot guarantee that AT&T's system, or any telecommunications system, will remain error-free. Bell Labs spokesman Karl Blesh says ''We're constantly working on software development to make the job more accurate. Our goal is to achieve no errors certainly, but we're also not going to delude anyone by saying it'll ever be one hundred percent perfect.''

The possibility that the crisis could happen again without warning doesn't seem to bother too many people. Even those who depend on the phones for their livelihood seem to simply take this all in stride as just another hazard of daily life in the modern age.

In Omaha, Chuck Waymire sits back in his office at the business he started twenty-two years ago, and contemplates. Waymire, who runs the Trans Mid-America Company with his wife and company vice president, Florence, has had to weather other calamities over the years. It has been one week since the long distance telephone service crash of January 15, 1990. He chuckles, and looks out his office window, surveying the green grass that is uncharacteristically showing its blades this winter.

"You know," he says matter-of-factly, "it's like a blizzard. It stops traffic, but then the blizzard goes, and it starts moving again."

Chapter Four

THE SPIRAL OF TECHNOLOGY

THE bottom line is that all software contains errors. Software errors are, sadly, an inevitable part of the software writing process, which is, after all, done by humans. The trick is trying to catch and fix the errors before the defective software ends up as a coded set of instructions inside an air traffic control computer or a medical radiation therapy machine.

It can't be done. Software can never be made completely error-free.

As software engineer Ware Myers notes in the journal *IEEE Computer* (vol. 20, November 1986), "The whole history of large, complex software development indicates that errors cannot be completely eliminated."

True, all software goes through an extensive testing process known as debugging. But, as any software developer will tell you, debugging cannot ever guarantee software will be error-free. Software can never be tested in all the real-life situations it will encounter. The possibilities are often nearly endless. So the odds are good that all software programs hold some errors that will not be discovered until they make themselves apparent in some real-world situation.

Software seems to suffer from four recurring problems. The program encounters a situation the programmer did not anticipate; the program is used for a purpose for which it was not designed; the programmer fails to consult enough with the person who will be using the program; or worse, the user goes back and fiddles with the program while it's still in the mid-stream of development.

AT&T is usually thought to have developed some of the most complicated, yet reliable software in the world. The software in its call switching network is among the most complex in use today. But AT&T officials are the first to admit their software is riddled with flaws, many of them hidden.

Officials from Bell Labs told members of Congress who were investigating software reliability the startling truth, "even the simplest software upgrade introduces serious errors. Despite our best efforts, the software that controls the telephone network has approximately one error for every thousand lines of code when it is initially incorporated into the system. Extensive testing and simulation cannot discover these errors."

The difficulty is that software itself is difficult to grasp. It cannot be laid out in an architect's blueprint or rendered in a clay model. Even a simple flow chart of a software program soon becomes an intricate maze of boxes and arrows that intersect in three dimensions, showing the various logic paths and decision boxes the computer can follow as it makes decisions. Flow charts diagramming large programs can easily fill a room as they display overlapping routes that data can follow as they weave their way through the computer.

Software is both invisible and unvisualizable to the mind's eye. It is abstract. Mike Champness, the software manager who helped develop the programs for the Air Force's F-16 fighter plane, explains, "You can't see it, you can't touch it, you can't put your arms around it. You can just sort of imagine it." One can only see what software *does*.

"Software programs," says University of North Carolina professor Frederick Brooks, a leading expert on computer programs, "are more complex for their size than perhaps any other human construct."

To a certain extent, software is tangible. It's made up of long strings of letters and numbers. But what it does and how it works remains only in the mind of its developer. "It's a very creative business," Champness explains. "It demands a lot of imagination."

The problems grow as software becomes more extensive. Software consultant and former Control Data software developer Ronald Enfield relates, "The more complicated software gets, the more errors there will be." The problem is not as simplistic as it sounds. As computer programs become larger, they grow in complexity at a rate greater than their growth in size. A program twice as big may be more than twice as complex.

A computer program quickly becomes an incredible mathematical trap. For example, a single tiny section removed from one computer program taking up a mere ten lines of code may have, say, three possible outcomes. Nothing too complicated. Put two of those small sections together to form a slightly larger program, and the number of possible outcomes triples to nine. By the time you get to six sections, the number of possible outcomes grows to 756. Put sixteen small sections together, and you suddenly have possible results that number more than forty million.

Testing each and every one of those possible results to make sure they all work correctly becomes impossible. By way of a real-world example, air traffic control computers use a software program that consists of sixteen thousand lines of code, which is relatively small by modern standards. But the results of that program are complex. How complex? The number of possible outcomes from that program is greater than the total number of atoms in the universe.

Such complexity also applies to the people who must write the programs. A small program, such as one for a video game, can be written by one person. But large programs have to written by teams of programmers, who somehow must share the common vision of what the program must do, and how it is to achieve that goal.

It is estimated that the Navy's new Sea Wolf attack submarine will require nine hundred programmers to write the software for its AN/BSY-2 combat computers. In a system that large and unwieldy, no one person can understand how the entire system works.

Having many programmers work on the same project runs counter to the fact that software development always has been a highly individualistic enterprise. As Champness relates, "You've got a guy with glasses and a little pocket protector who hangs out in this darkened room, and he just sort of creates things. Well, that's fine when you've got a small program. But when you've got a large program that interrelates with other programs, they don't play well together. Suddenly you don't know what one guy is doing or what another guy is doing. It goes against the grain of how software has come about."

Software development is a creative process, and what each programmer does affects every other programmer on the project. Trying to rush development of a software program is like telling Michelangelo to hurry up and finish the ceiling of the Sistine Chapel. Now, imagine making changes to the Sistine Chapel design when it's already half-finished, and making sure the changes are coordinated amongst a team of nine hundred separate Michelangelos. ("And oh, by the way, Mike. Can you make the thing a little brighter? The whole thing looks kinda drab, don't you think? And go back and fix the smiles on all those people, would you? They look too happy.")

Consider first how software programs develop. The people who will be using the software, whether air traffic controllers, business people, doctors, or whatever, describe the requirements of what they would like the program to do. As Enfield picks up the process, "You proceed to a design, and that goes through exhaustive design review, where everybody looks at it and tries to poke holes in it logically. And then it goes to a program development, where they write the programs. And then they review the code step-by-step, from the very general to the most specific to see if it reflects what they wanted it to do. But in each part of the process generally they do find errors, and they have to go back in and correct it. As you correct errors, you need to recheck the design and the requirements to make sure you haven't strayed."

As Mike Champness explains, "There's always a bit of a shot

in the dark when you're doing something like that. When you do that you in effect need to go through the whole top-down process again with this little change to make sure the ground rules still fit.'' An engineer who is designing a bridge can always go back and do a stress analysis to see how a change will affect the entire structure. A software engineer has no such luxury.

The problems grow worse if customers or those using the software come back in the middle of development and demand changes to the program. Changes in software usually require that the programmers go backwards through the entire programming process and essentially start over again. This is necessary because a change in even one part of the software will often affect how the entire program works, forcing a nearly complete redesign and more testing.

Champness complains, ''They'll say it's only one small change, right? Well, you know, one small change becomes two small changes, then ten small changes. They seem incapable of keeping the same requirements throughout. They don't realize it takes a lot of time and effort.''

The temptation of end users to make new demands of software is one of the biggest problems that generates errors. Because software is by its very nature, malleable—after all, it's only made up of a string of numbers and letters—people often assume a change in a software program is an easy thing.

As the Congressional Office of Technology Assessment notes, ''No one would ask a bridge builder to change his design from a suspension to an arch-supported bridge after the bridge was half-built. [But] the equivalent is often demanded of software builders.''

Changes further tax the ability of programmers to eliminate errors. Past a certain point, frequent changes begin generating new errors faster than they can be removed.

One would hope that testing would eliminate all these errors. But unfortunately that's not the case. Initial testing of a new piece of software invariably reveals a significant number of errors. Software designers then go back and make changes to correct the error. The process greatly reduces the number of errors, but never reduces them all the way to zero. Given a common industry average of one error for every thousand lines of code, a software program of one million lines of code would contain one thousand errors. If testing were to correct ninety percent, that would still leave one hundred errors in the program.

Testing provides no guarantees the corrected program is then error-free. The ground processing system software for the space shuttle underwent over two thousand hours of simulation testing before the first shuttle ever took off. The software was deemed "stable," meaning it performed its functions reliably enough to be used. The testing had uncovered more than two hundred errors. But on the first shuttle flight, twenty-four more errors were found, four of which were judged to be "critical" or "major."

Worse, correcting an error can, and often does, introduce a new error into the system, creating a computing Catch-22. For that reason, software errors are sometimes left uncorrected. Correcting it would simply carry too great a risk of creating new errors. Software users are often simply given bulletins, warning them about known software errors.

It's interesting to note that software programs almost never carry warranties. Instead, they usually come with disclaimers relieving the programmer of any responsibility or liability for any problems that develop because of errors.

And the potential consequences of those errors are also unpredictable. Enfield notes, "Hopefully you've found and fixed the most serious bugs. But instead of just misspelling a word on the screen, there's always a chance the next bug you find may cause the computer to crash."

Software programs must be unerringly accurate. Even simple typographical errors can ruin a program. Typing a semicolon instead of a colon, or simply misspelling a word can mean disaster. On July 22, 1962, NASA had to destroy its *Mariner 1* space probe because of a missing hyphen in one of its software programs. The missile had just taken off from Cape Canaveral when it began behaving erratically. Range safety officers destroyed the rocket before it could endanger populated areas. Instead of ending up over Venus, *Mariner 1* rested at the bottom of the Atlantic.

A minor equipment malfunction had unearthed a software error in the guidance program, which had previously worked reliably on several missions. The cost to NASA: $18 million. The goof has since been referred to as "the most expensive hyphen in history."

A computer is a marvelous piece of machinery. It will follow instructions with uncanny and repeatable precision. But those instructions must be equally precise. Unlike a human, a computer cannot look at a misspelled word and guess what it thinks you meant by looking at the context in which it was used. As computer programmers say, garbage in, garbage out.

How difficult then is it to debug a program that's one million lines of code long? Picture this: the Manhattan *White Pages* telephone book contains about one million entries. Imagine that after you finish writing it, you realize there's an error in there. Somewhere. Painstakingly, you go over every entry. You double-check it. What do you think the odds are everybody's name, address, and phone number are listed correctly?

Errors in software are considerably more complicated, because each line both affects and is affected by other sections of the program. A computer software malfunction may give you some vague idea where the error occurred. But pinpointing the exact source of the error often means tracking carefully through endless codes and logic paths.

One computer programmer with DuPont complains in *Business Week* (no. 3051, May 9, 1988), "I might write two hundred lines of code on Monday, and if there was a flaw in it, spend all day Tuesday trying to figure out what was wrong."

Then consider that the software for the space shuttle holds over *twenty-five million* lines of code.

There are no easy answers on how to improve software development. The ability of computers to handle tasks has improved by leaps and bounds, increasing an impressive five-hundred-fold in the last thirty years. But while computer hardware technology continues to grow by leaps and bounds, software programs are still written the same way they were forty years ago in the dawn of the computer age. Huge teams of programmers laboriously hack out perhaps six lines of code per person every hour. To write a one million line program is a daunting task to say the least. Robert Terderman of American Express described it in *Business Week* (no. 3051, May 9, 1988) as, "akin to building a 747 jet using stone knives." A Congressional Office of Technology Assessment report glumly notes, "Advances in computing hardware have come rapidly, but software development is notoriously slow and costly. The nature of software causes progress to be slow, and there is no prospect for making a radical change in that nature."

Clearly, new technologies must be developed to eliminate, or at least reduce, programming errors.

One potential solution is to "teach" computers to spot errors and then work around them. In such a "fault-tolerant" system, a series of computers monitor each other and "vote" on how to handle the crisis. The space shuttle, for instance, uses a highly advanced fault-tolerant architecture. Four identical computers, using identical software programming, operate all critical flight operations. Each time a decision is needed, the computers vote. If one of the computers disagrees or has a fault, the other computers vote to ignore it. If two more computers fail, control of the shuttle is handed over to a fifth "standby" computer that uses a completely separate set of software commands.

On occasion, though, the increasing complexity of fault-tolerance can create problems of its own. As the first space shuttle prepared for launch, the different software programs in the main and back-up computers could not "agree" on how to synchronize their count-down clocks. Each concluded the other was not operating properly. The launch was scrubbed.

Even a triple-redundant system couldn't save the Toronto Stock Exchange from crashing in August 1989. The exchange was shut down for three hours, forcing trades to be switched to the Montreal Exchange. Three different computer systems went down. An official from Tandem Computers, which designed the system, complained to the *Toronto Globe & Mail* (vol. 145, August 21, 1989), "I've never heard of three failures at one time. I don't know what the gods are doing to us."

Not only is fault-tolerance no guarantee of freedom from problems, it is also extremely expensive. In essence, you're building the same system three, four, or even five times. And fault-tolerant systems, because of the constant checking for errors, are larger and generally run more slowly than competing systems.

Better answers lie in developing methods of writing software that reduce errors. "Structured programming techniques" is becoming the latest buzzword phrase in software development. As programs become more complex, they need to be developed in stages. Each stage needs to achieve its own goals. If the goals are not met, software writers need to go back and rework the program before proceeding on to the next step. This helps reduce the tendency to say, "We'll fix it later as we go."

Similarly, some software developers are already using a technique called "clean room," named after the dust-free environment used to make computer chips and other semi-conductor components. Development of a software program is broken down into increments. At the conclusion of each incremental development, the program is statistically tested to prove the product is producing the expected result.

Another solution is a process called IV&V, Independent Verification and Validation. Someone who is not connected with the development of the particular software comes in and doublechecks its design. Independent tests are run at various stages of development to make sure the software works the way it's intended.

Asst. U.S. Comptroller General Ralph Carlone, who has investigated troubles in numerous government computer projects, stated in a report that "an independent verification and validation process is important to reduce risk in software development."

Structured programming, clean-room techniques, and independent verification and validation will not guarantee error-free software. These methods are better at simply making sure the software does what the user wants it to. But a combination of the techniques would go a long way towards reducing software errors.

As Mike Champness puts it, ''Structured programming techniques are basically the way to develop software. But it demands a lot of discipline. You've got to start from day one with good firm requirements and good development procedures. Then, too, you need independent test people who didn't develop the software but understand it.''

In the long term, new programming techniques are on the horizon that promise to drastically reduce software errors. Ironically, the answer lies in using more computers. Software developers in Britain and France decided to use computer-aided engineering programs in software development. The thinking was, if computers can design bridges or airplanes, why can't they be used to design other software programs?

The technique is called Computer-Aided Software Engineering or CASE. The tasks assigned to a program are broken down into sections. These pre-fabricated sections of software code can then be assembled to form complete programs. One can imagine that CASE is to software development what the assembly line was for the Model *T* Ford. Instead of writing out a software program line-by-line, CASE uses interchangeable parts to standardize software.

Programmers often find that they are simply re-inventing the wheel each time they write a program. Studies have found that only fifteen percent of the software code in a program is new. Most of a supposedly new program is taken up by code that has already been developed for other applications and uses.

Besides making software writing quicker, CASE also makes for fewer errors. Many software errors are typographical mistakes and misspellings a programmer makes. Automating the writing of software eliminates those errors. Since each section has already been thoroughly tested, the resulting program is likely to be far less prone to software errors.

When errors do occur, they are often easier to trace and fix. As

Mike Wybo of the University of Minnesota's Management Information Systems Research Center says, "It's like in your car. If something goes wrong, you can just pull out the bad module and snap in a new one instead of trying to trace through a whole bunch of code to try and figure out where the problem is."

Companies have found the CASE technology can greatly cut the cost of debugging programs. When Raytheon Corporation unveiled its modular business software development system, called ReadyCode, it cut the cost of programming sixty percent and greatly simplified design and testing of new programs.

But many companies have been reluctant to use CASE techniques. Part of the reason is inertia, the "we've always done it this way" philosophy, and part is the high cost. CASE can cost upwards of fifteen thousand dollars per programmer, and for companies with hundreds of programmers, that is clearly an enormous expense, one that is difficult to justify for many cash-strapped corporations.

Use of CASE techniques took a massive step forward when IBM stepped into the ring in late 1990 with a new product called AD/Cycle. The product features reuseable software modules to assist software programmers in building complex programs.

Being the undisputed world leader in "mainframe" computers as well as PCs, or personal computers, IBM wields marketing influence second to none. A popular saying in the computer industry is that "when IBM sneezes, other computer companies get blown away." Therefore, IBM's introduction of CASE was seen by many as CASE technology's coming of age. Jim Webber is a computer consultant for sixty major corporations, including AT&T, Prudential Insurance, and Warner-Lambert Pharmaceuticals. He notes IBM's move "creates an acceptability out there that wasn't there before. They kind of legitimatize [CASE in] the marketplace."

CASE is not without its drawbacks. Because its parts have to be adaptable for a myriad of different roles, CASE programs can often be large and unwieldy. They will then run slower on most computers. And it's difficult to develop very complex or specialized functions that way.

Use of CASE techniques to develop new software programs greatly reduces software errors, but can never eliminate them, especially as programs grow ever larger and more complex. Even CASE programs are, after all, designed by humans.

And the problem of software being used in ways for which it was not designed or in ways unanticipated by programmers can never be eliminated. As software consultant Ron Enfield says, "You can design some mathematical functions to be very, very reliable. But when you deal with people, or when you deal with events in the real world, things can always happen in ways you didn't anticipate. So the risk of things going wrong is there, and it is increasing in many, many different ways."

Dr. Peter Neumann, a leading software expert at SRI International, a computer think-tank, cautions in *Science News* (vol. 130, September 13, 1986), "Experience shows that even the most carefully designed systems may have serious flaws. In addition, human misuse can compromise even the best systems. Even if we are extremely cautious and lucky, we must anticipate the occurrence of serious catastrophes in the future."

As the Congressional Office of Technology Assessment somberly concluded, "Errors in large computer programs are the rule rather than the exception. Errors in the code, and unforeseen and undesired situations are inevitable."

Chapter Five

MONEY, MONEY, MONEY

Ask the CEO of any major corporation about software errors, and he or she will probably grimace and tell you all about them. Any company of reasonable size has faced them, and chances are they have cost the company millions of dollars in losses.

In the world of big business, software errors mean big money. Even small errors in code can be extremely costly. And guess who usually ends up paying for it in the end?

Companies are consistently unable to develop large software projects without going tremendously overbudget, often by several million dollars. Delivery of the software is often late, and even then, the software does not work properly.

The failure to grasp how software is designed produces a nightmarish array of problems. Software for commerce becomes an oft-repeated exercise in wasted money and trouble-plagued projects. And it is often the consumers and stockholders who end up ultimately footing the bill for the painful lessons learned. The examples are both numerous and costly.

The gleaming skyscraper stands as an imposing symbol of Bank of America's power, towering over downtown San Francisco, looking down on its lesser rivals. But inside the company's fifty-two-story world headquarters, the use of the word MasterNet is practically forbidden. To this day, Bank of America will not even admit the computer system ever existed. But, privately, one company official admits, "We had some problems with it."

The Institutional Trust Services Department is one of the most prestigious and important divisions of the nation's third-largest bank. Each year, it handles more than $38 billion from some of the nation's largest pension funds and other institutions, which retain Bank of America to find the safest and most lucrative investments. With the historic bull market in stocks still in its early days, Bank of America officials felt the time was ripe to aggressively go after new business.

So in 1983 the company began drawing up plans to replace its computer systems that handled institutional accounts. Company officials said the investment was needed for the bank to remain competitive in the growing global market for financial services. A small, but highly respected company in Pennsylvania was recruited to develop the new software. Its president had been one of the pioneers in developing software for managing trust accounts. Within two years, the Pennsylvania company promised, Bank of America would be handling its institutional accounts with speed and efficiency unmatched in the industry.

The shakedown period for the new software proved troublesome. The program refused to work properly on Bank of America's computers. Bank officials quickly grew frustrated. Several rival banks were already using new computer systems. In March 1987, the bank decided to simply take the plunge: its entire system was converted to MasterNet. The new program crashed.

MasterNet went down for days at a stretch. Interest payments were not credited to accounts on time. Transaction records failed to show if securities had been purchased from other Bank of America accounts or from outside the bank. Customer statements fell months behind. Auditors had to be called in to double-check every transaction by hand. Insiders estimate the MasterNet fiasco cost Bank of America at least $1.5 billion in business, as institutional investors, frustrated by statements that were months late, pulled out their stakes and took the business elsewhere. The costs didn't end there, either. Unable to make the system work, Bank of America threw in the towel in January 1988. The huge MasterNet program was abandoned, after Bank of America had spent $20 million on it. One company worker predicted, "Heads are going to roll." They were right. Both the head of data processing and the head of the trust division were forced to resign. The bank's largest accounts were turned over to a bank in Boston. MasterNet cost Bank of America at least $60 million in overtime expenses and late interest charges.

Jettisoning MasterNet, painful though it was, provided only temporary relief. Bank of America quickly became entangled in another, even larger software mess, though not of its own making this time.

United Education and Software managed over $1 billion in federally guaranteed student loans, collecting payments from student borrowers under contracts with several major banks. Among its many clients, Bank of America. In October 1987, United Education and Software switched its computers over to a new software program. The new program contained some major errors, and quickly began running amuck.

Students who were still in school began receiving delinquency notices, even though they weren't scheduled to begin making payments. People who were behind on their payments were never notified. Others who had been faithfully making payments began receiving notices stating they were behind. Students who sent in checks had their payments rejected. Payments were improperly credited. United Education president Aaron Cohen said he was aware the software was troublesome, but thought it was just routine teething pains common to any new program.

A Department of Education audit revealed the true extent of the damage: the eight months that the program ran wild could end up costing Bank of America and the several other major international banks who used United Education up to $650 million. The federal government says it won't honor its guarantees on the loans, and the banks may have to default on them.

And Bank of America is far from alone.

Software has now become one of corporate America's biggest costs. U.S. companies will spend over $100 billion this year on software programs, an amount nearly equal to the annual gross national product of Sweden. United Airlines spends $120 million a year on software for its Apollo reservations computer. Merrill Lynch shells out an estimated $160 million yearly for its financial analysis and customer account software.

But the corporate scene is littered with scores of overdue, troublesome, and tremendously overbudget software projects that are costing corporations and ultimately, consumers and shareholders, millions of dollars. And a software manager for a top accounting firm says, "We're still not learning our lessons from our mistakes." The list of companies experiencing troubled software programs continues to grow alarmingly.

Allstate Insurance, the nation's largest insurance underwriter, had grand plans when it decided to revamp its computer operations. The project was unveiled in 1982. It still isn't finished.

The company wanted to set a new standard for office automation. All paperwork inside the sprawling company would be computerized. The time needed to develop new types of policies would be shortened from three years to just one month, giving Allstate a tremendous competitive advantage and flexibility in tailoring its coverages to changing lifestyles and business trends.

Less than a year after the company began the project, things began to slowly go awry. Deadlines for completion of various stages of the project came and went. The cost mushroomed from $8 million to $15 million. Finally, Allstate fired the system developer and hired a different firm to complete the project. The estimated cost: $100 million.

The Bank of America and Allstate stories may be extreme examples, but they are by no means isolated ones. Allstate's new consultant, James Willbern of the accounting firm KPMG Peat Marwick admits, the industry's "ability to install systems is pretty crummy." The problem of software development running incredibly overbudget and then not working correctly is more than simply widespread; among large corporations it is practically universal.

KPMG Peat Marwick surveyed companies and found thirty percent admitting to having fallen victim to ''software system runaway projects.'' Among Fortune 200 companies, Willbern found there are basically only three types of companies: those which have had expensive runaways, those which are currently experiencing them, and both. In the United Kingdom, it is estimated software problems cost businesses nearly $1 billion a year. A regional U.S. telephone company watched the projected cost of a software project double to a whopping $700 million. As Willbern says, ''It can cause some companies to go under, to go bankrupt. They are totally dependent on computer systems.''

Part of this abysmal record is due to the nature of the beast. Business software programs are becoming extremely complicated. A computer is not like a television set—you don't just buy it and turn it on. Especially for large businesses, systems must be painstakingly tailored for particular tasks. In essence, you are buying a huge, custom-designed software package. The simple business spreadsheet programs that became popular in the 1970s contained perhaps five thousand lines of software code. It's not uncommon for today's business programs to have lines of code numbering in the millions.

Robert Nagel, vice president of a software development firm, Cognition Inc., says, ''Today's business software programs are considerably more complex. The computing power of today's computers has quintupled in five years. And as people find newer and newer uses, newer and newer problems they can solve, the programs are becoming extremely complicated.''

Computers are now indispensable for modern businesses, and no longer just valuable tools. The personal computer is becoming more powerful, and now they're starting to become a standard piece of business equipment on everybody's desk. Virtually every business, down to the local mom-and-pop grocery store uses computers to run their operations.

And companies now want computers to control all facets of their far-flung operations, including inventory, data processing, order processing, and distribution. A University of Minnesota study found that if the software were to completely fail, banks would have to close within two days, retail distributors would be backlogged within four days, and factories and other businesses would have to close by the end of one week.

Even a medium-sized company would find its losses begin to snowball, until by the fourth week the company would lose $2 million. Despite such daunting risks, companies continue to press ahead with ever-more ambitious computer projects.

In 1983, Blue Cross and Blue Shield of Wisconsin spent $200 million to consolidate and streamline its computer operations. The new computer software wound up sending out more than $60 million in overpayments and duplicate checks. When a data-entry clerk answered a question by entering ''None'' into the system, the computer promptly sent hundreds of payment checks to the nonexistent town of None, Wisconsin. By the time it was all straightened out, thirty-five thousand policyholders had switched to other insurers largely because of frustration over such errors.

Such problems can inflict staggering amounts of damage. The Bank of New York learned all about the cost of software errors the hard way in November 1985. At the end of one particular trading day, bank officers were shocked to discover the bank somehow was an incredible $23.6 billion overdrawn. The flaw was in a complex software program that handled sales of government securities. Such securities are available in a wide variety of maturity dates, prices, and interest rates. The software was designed to handle up to thirty-six thousand different types of securities.

But a surge in trading activity on that day triggered a flaw in the software that had until that point gone undetected. The error caused Bank of New York's computer to stop transferring securities to other banks. At one point in the day, Bank of New York owed other banks as much as $32 billion.

Because by law banks are required to balance their books, Bank of New York had to borrow $24 billion overnight from the Federal

Reserve, by far the largest such emergency loan ever made by the Fed. The bank was forced to pledge all of its assets as collateral. The mess took two days to straighten out. The interest on the loan was $5 million.

Such problems are, surprisingly, not that uncommon in banking. The Federal Reserve Bank of New York reports that software errors and computer failures caused banks to have problems settling their books at the end of the day more than seventy times in 1985 alone.

And the amount of money handled by computers in the business world is truly amazing. Up to $2 *trillion* is transferred by computer between banks and securities firms every day. That total is more than a quarter of the nation's annual economic output of goods and services.

Given the high speed of these computer transactions, software errors can scramble mind-boggling amounts of money in just minutes. A number of banks in the United Kingdom use a computerized transfer system known by its ever-so-British acronym, CHAPS: the Clearing House Automated Payment System. But in October 1989, CHAPS turned out to be a far less than agreeable fellow.

CHAPS contained an undetected software error that caused certain payments to be duplicated. On this particular day, CHAPS mistakenly transferred two billion British pounds (then about $3.4 billion U.S.) from one of its member banks to various customers in the United States and the United Kingdom.

From start to finish, the CHAPS multi-billion dollar spending spree took less than a half hour.

CHAPS officials were nervous as they awaited the return of the excess money, and for a good reason: the transfers are guaranteed payments, and legally, the money is irretrievable. Fortunately for CHAPS, owing to the goodwill of the close-knit banking community and its corporate customers, all of the money was subsequently returned to the rightful accounts.

But despite such potential (and costly) pitfalls, today's twenty-four-hour international trading markets leave little choice but to rely on computers. As then-Federal Reserve chair Paul Volker told a

House hearing, "Like it or not, computers and their software systems, with the possibility of mechanical or human failure, are an integral part of the payments mechanism. The scale and speed of transactions permit no other approach."

Competitive pressures leave little choice but to computerize. The business environment is rapidly becoming more complex. Companies that once may have done business only in Connecticut now find themselves dealing with customers around the world. Financial companies are tied into global financial markets that are open twenty-four hours a day. As Merrill Lynch's director of investment banking systems Joseph Freitas told *Business Week* (no. 3179, November 26, 1990), "Our survival depends on fast information and knowing how to analyze it."

Companies are pushing their computer departments to develop software that will be a "magic sword" delivering a competitive advantage. Bill Nance of the University of Minnesota's Management Information Systems Department explains, "There are a lot of examples in history, of companies using computer technology to gain competitive advantages, where they've managed to change the face of the entire industry."

The best-known example may be computer-driven program trading in securities markets. High-speed computers take advantage of minute, fleeting differences between stock prices and values of stock futures contracts. The programs will buy and sell huge volumes of shares and contracts on the basis of differences in price, rather than whether a particular stock is a sound investment.

Such high-speed trading can whipsaw the stock market a hundred points or more in a matter of minutes. The impact on the financial markets has been enormous. Analysts agree program trading has greatly contributed to the extreme daily volatility of the stock market. The New York Stock Exchange now has special rules that suspend computerized program trading when the stock market rises or falls too rapidly. The technology has also contributed to the flight of the small investor from the stock market. Much of the general public no longer believes it can compete with sophisticated trading programs. A recent *Business Week*-Lou Harris poll found that seventy-nine percent of the public had "hardly any" or "only some" confidence in the people who run the stock market. One of the major reasons cited was computerized trading of stocks.

While computer programmers are under tremendous pressure to develop elaborate new programs that will give the end users competitive advantages, they are also under tremendous pressure to make the programs simpler. Twenty years ago, when only a handful of a company's employees might actually get their hands on a computer, a company could afford elaborate training programs for the select few.

But now that desk-top computers and workstations have become nearly as common as typewriters and telephones, the pressure is on to make software simple to use. "People are looking for programs they can learn in a few hours, and be solving problems by the end of the first day," sighs an exasperated Robert Nagel, a software corporation executive. "It makes it very hard from a software development point of view to create programs like that."

In truth, the marriage between business people and computer programmers has always been an uneasy one at best. For starters, the two worlds are inhabited by people who barely even speak the same language. Jon Laberee is systems manager for Norwest, a major Midwest bank. "The banking people on your trading desk will explain what they need and be talking about swaps, caps, floors, and stuff like that, and the programmer doesn't know what that is. And then you talk to the programmer and they're talking bits, bites, and nanoseconds. Nobody knows what anybody's talking about. You really almost have to teach someone your business before they can go ahead and write the software that's specific to you. The industry is really looking for people who are power programmers but also understand business. That's very hard to find."

Laberee complains that the way most college curricula are set up, business students may take only one or two basic computing courses. Computer programmers often never take business courses, although they may later be called on to design huge business software systems.

As computer consultant Michael Hammer explained to *Fortune* (vol. 120, September 25, 1989), "Imagine putting together a manual on how to assemble a car for someone who had never even seen one. That's what you're doing when you write a program."

Problems of miscommunication become worse in designing software to suit broad needs within a very large corporation. People within the same management levels often don't speak the same language. Different departments have vastly different needs, and trying to design software to encompass everyone's needs can be nearly impossible. Bill Nance of the University of Minnesota, has analyzed why companies have such difficulty developing software. "A customer to the marketing department is not the same thing as the way the finance department, for example, views a customer," says Nance. "Your marketing people need sales history and demographic information: what the customer has purchased over the last ten years, where they live, etc. Your finance people need credit histories, payment structures, credit worthiness. The problem becomes how can we design the system to serve everybody's different needs?"

To make matters worse, software developers often don't consult with the very people they are supposedly building the system to serve. Max Hopper has seen more than his share of computer disasters, having worked as head of information services for both Bank of America and American Airlines. He notes in *Fortune* (vol. 120, September 25, 1989), "Too often the high priests in the centralized data processing departments try to come up with new systems without including the business people in the process." Management often does not get deeply involved in software development because they are frightened by the technical nature of it. The software that results then often fails to take into account what the needs are of the people who will be using the program.

And the strains being placed upon modern business programmers are enormous. A programmer for a bank automated teller machine program, for instance, must make certain that deposits or withdrawals made at a twenty-four-hour teller machine in Hawaii are, with perfect accuracy, credited to the proper account in banks in New York or Chicago. At the same time, the programmer must ensure that the system is guarded against computer hackers who may attempt to disrupt or alter computer records or even steal from accounts.

Given such complicated requirements, it is little surprise there are occasional glitches, some more serious than others. In March 1990, an automated bank cash machine in Oslo, Norway suddenly began handing out ten times more money than customers were requesting. The Kredittkasssen Bank suddenly found itself with a kind-hearted computer on its hands. Customers who requested one-hundred-kroner denominations were instead handed one-thousand-kroner bills by the generous machine.

Fortunately for the bank, the largesse was short-lived. Honest patrons promptly reported the software-inspired giveaway to police who closed down the faulty cash machine. The computer at least managed to keep a record of its impetuous generosity, and the cash was subtracted from the accounts of the lucky customers.

Westpac Bank of Australia was not nearly so lucky when problems struck in May 1987. When a new software program was installed for the computers controlling their ATMs, customers soon discovered they could use their bank cards to withdraw more than the daily allowed maximum and could even withdraw more money than they had in their accounts.

When Westpac officials finally discovered the problem, they shut down all of their ATMs throughout Australia. Bill Paget, Westpac's electronic banking manager, estimated the bank had lost several hundred thousand Australian dollars. But because of the high cost of recovering the money given out, Paget announced the bank probably would not take legal action to get its money back. Westpac blamed insufficient testing of its new software.

Ironically, while businesses rush development of new programs to gain fleeting competitive advantages, software development becomes comparatively slower and slower as the programs become increasingly complex. Although the power and speed of business computers doubles every three years, software productivity rises a mere four percent a year.

The average development time for new business software programs has climbed from six months to thirty months in the last fifteen years. The average new business software program takes thirty-two thousand workdays to write. It would take a team of thirty-six programmers almost three years.

In a rapidly changing business environment, such drawn-out timetables often mean software is obsolete before it can even be completed. "It takes so long to build these things," says University of Minnesota information system analyst Mike Wybo, "you specify the system you want today, and by the time they come back five years later, you're not in this job anymore, the corporate environment has changed, and suddenly you're not interested in those kinds of problems anymore, and you need the system in a different format."

But changing a software system that's under development is fraught with hazards all its own. Software developers tend to agree that businesses often underestimate how easy it is to redesign software as their "wish list" of features and needs evolve. "People tend to underestimate the cost of changes," says one developer. As a new software program is born, "as you go from one step to the next, the cost goes up to make the change. Depending on when you request the change, it can be inexpensive or it can be extremely expensive." Changing a program that is nearly finished can cost up to one hundred times more than it would cost to make the same change if the software was still in the initial design stages.

Worse, as business software grows in complexity, it becomes nearly impossible to test the software properly in order to catch potential errors. Programs can no longer be tested in every conceivable real-world scenario. Computer specialist Mike Wybo points out, "The more complex a program is, the more difficult it is to test all the possible combinations of values and parameters. And so you have a higher probability that something will go wrong with it, or that someone will uncover something in the future."

To try and ferret out errors before a particular piece of software gets put into widespread use, many software development firms now engage in a procedure called beta testing. Sample versions of the software are sent out to companies. The customers are urged to run production-level problems using the software to try and uncover errors in the software. But beta testing can take up to a year. In the rush to bring systems on line and gain a competitive edge, companies often decide to make do with only limited amounts of testing.

KPMG Peat Marwick's Jim Willbern concludes, "I'd say it's pretty common. As you approach the 'drop-dead' date when a program's supposed to be finished, they often short-cut the testing." And when software development runs late, testing, since it's the final phase of a project, is often the first to be jettisoned.

IN September 1988, American Airlines discovered a software error had cost the airline as much as $50 million.

As accountants went over the airline's second quarter financial results they noticed that despite near-record profits, the airline was suddenly selling far fewer discount tickets. Closer analysis also revealed the airline suddenly had more empty seats than in previous years. The problem was traced to new computer software in American's Sabre reservations computers. Sabre is the world's most widely used travel reservations system.

Just months earlier, American developed its own "magic sword": a new type of software called "yield-management" programs. The programs revolutionized the airline industry. Yield-management programs determine how many discount fares, and of what type, will be allotted to each flight. A heavily travelled route, for instance, would receive few discount seats, in order to maximize revenue. A less popular route would receive a higher proportion of discounted seats to try and woo more passengers on to the route. The goal was to maximize profits on each flight, and the programs would later prove hugely successful, generating millions of dollars in new revenue. They have since been copied by hotels, rental car agencies, time-sharing condominium developers, and other airlines.

But when the yield-management programs were first used, the Sabre computers showed that certain flights were sold out of discount fares, even when such seats were still available. When customers called for reservations and requested a discount fare, they were told there were none available, and were referred to the competition.

American officials reluctantly admitted the problems were the result of a new software program that had not been fully tested. That American had rushed the programs into the system was a bit of a surprise. The airline is admired within the reservations industry as the leader in technology. When American built its state-of-the-art reservation center in 1987, the company took enormous precautions to ensure the integrity of its computers. The huge Tulsa complex uses eight separate IBM computers to try and ensure that no single failure can knock out the entire system.

Consultants had even put up elaborate defenses against human interlopers and Mother Nature. The entire complex was constructed sixteen feet below the ground, and was reinforced to withstand the ravages of a direct hit from a 350-mile-an-hour wind tornado. The center holds a three-day supply of food and 125,000 gallons of water. Standby power generators and 125,000 gallons of diesel fuel ensure the computers could run for three days without outside power.

Security is worthy of the Pentagon. To gain entrance, employees are required to pass through a bulletproof, steel-enclosed booth. Inside, a retinal scanner checks the distinctive ''fingerprint'' of blood vessels in the back of each employee's eye, and matches it against the employee's file before granting entrance.

When it was completed, director George Maulsby noted proudly (*Newsweek*, vol. 110, December 28, 1987), ''We can't afford fires, natural calamities, or some crank on a rampage.''

But all the precautions meant nothing when confronted with the far more subtle threat of highly destructive software errors.

Less than a year after the yield-management error was corrected, on an otherwise uneventful day in May 1989, technicians at American were installing new memory disk drives into the Sabre system.

The installation triggered a software error that erased some of the information in the memories. The airline could no longer access information as to which passengers were booked on which flights.

The airline's elaborate defenses were in vain as the software error jumped in rapid succession form computer to computer. American officials said afterwards they couldn't have done a better job of disabling Sabre if they'd set out to do so deliberately. "It was our worst nightmare," one vice president lamented.

Fourteen thousand travel agents around the country had to make phone calls to book flights, instead of doing it by computer, and tickets had to be written by hand. A tall order, considering American has over twenty-three hundred flights a day, carrying a total of more than a quarter of a million passengers. The shutdown was solved twelve hours later. American would not discuss how much revenue had been lost because of the error.

Victimized by such errors, weary over delays and cost overruns in new software packages, and the difficulty of making changes, many companies are now resorting to guerrilla warfare tactics in dealing with their software needs. University of Minnesota computer analyst Mike Wybo notes a trend towards smaller, more adaptable software. The new software is custom-tailored to the needs of each department in a firm. Separate departments are developing "cardboard" or throwaway software packages. Such systems offer a great deal of flexibility. They can be developed quickly to deal with specific needs or problems, and then thrown away when they're done, or when the program is upgraded.

Citibank, for example, recently split off its data processing department. It now consists of thirty-nine separate divisions, each one tailored to the particular needs of a corporate department.

Merrill Lynch subdivided its data processing down even further. The giant investment firm simply scrapped its giant IBM computer and, instead, now relies on an army of eleven hundred personal computers, scattered throughout offices in thirteen cities. The computers are all tied together by a network, allowing exchange of programs and information.

Although such moves can cut costs and save time, the trend towards cardboard software programs is a potential timebomb. As each department in a business heads off in its own direction for computing needs, it creates a computing Tower of Babel, where departments can no longer compare notes and coordinate strategy. "It creates terrible problems," Wybo warns. "Like incompatible data between departments, a lot of redundant data floating around. Someday you're going to have five hundred different systems in a company all calculating the same thing and an organization coming up with five hundred slightly different answers."

And rarely is there such an animal as a "simple" or even a "typical" business transaction, especially in finance. A business loan can be done on a 365-day year, or a 360-day year. The loan terms could call for interest rates to be updated every month, using one of several different accrual methods. Even a simple factor such as the occurrence of a leap year can throw off a software program. The Pennsylvania Lottery discovered this when, in their first lottery in 1990, the Wild Card Lotto computers could not find the winning tickets. The simple rollover to the new year had thrown off the program.

Another problem is that software often magnifies human errors. The original version of the popular Microsoft MS-DOS system, which runs on IBM PCs, could erase memory disks with one keystroke. Many businesspeople watched helplessly as months, or even years of work vanished into an electronic oblivion.

A Florida construction company used the popular Symphony spreadsheet to calculate a bid on an office building project. The company controller misplaced $254,000 in costs that should have been charged to the $3 million project. Company officials won the bid, only to discover in horror that they now stood to lose hundreds of thousands of dollars on the project. The construction company went back to the developer and managed to rewrite the contract in time.

Larry Nipon, president of the Albert Nipon clothing manufacturer, caught an error that overstated profits by $1.5 million. He had made a simple error in using a program that calculates projected sales for the coming year.

Such examples pale in comparison to the Bank of Americas, Allstates, Blue Cross and Blue Shields, and American Airlines of the world. But all point up the dangers and frustrations of increased use of software in the business world. Even if they are designed with the utmost of care (and usually they are not), software programs are always in danger of running into situations that programmers did not anticipate.

When General Motors christened its gleaming new Buick City factory in Flint, Michigan, the company had high hopes. The plant was intended to be a showcase for advanced automotive technology. GM had sunk $400 million into the facility, purchasing some of the world's most advanced computer-controlled robots for performing assembly line work.

The plant, GM boasted, would produce the highest-quality cars at a lower cost than conventional assembly lines. But when the factory cranked up production, officials were horrified to see cars rolling down the assembly line minus windshields. Robots were used to install the windshields, but the visual sensors had not been correctly programmed to allow them to ''see'' black cars. The robots then blithely ignored those cars and sent them on down the line without performing any work on them.

It is worth noting, though, that the error has since been corrected, and Buick City, as General Motors had hoped, now produces some of the highest-quality cars in the world, equalling, and in many cases surpassing many of the leading Japanese and German auto-makers.

Slowly, companies are learning the hard way. The difficulties of business software development are much more managerial in nature than technical. Companies simply need to tackle complex software projects in smaller, more easily digestible bites. James Willbern of KPMG Peat Marwick notes companies can save themselves enormous headaches merely by keeping the size of the project down. Willbern has no shortage of demand for his services. In their first two years, he and his team of thirty-five specialists billed companies for more than $30 million in return for helping them rein in their software problems.

As a software development project expands to involve eighty people or so, about thirty percent of the time is spent simply communicating between people. When it goes above eighty people, Willbern says, "It just grows tremendously. You spend too much time just communicating with everybody that needs to know, plus everybody that wants to know about how it's coming. We try to keep projects down to a certain size."

Software development needs to be checked every few months to make sure the project is on track. If the goals are not being met, development should be suspended until corrections are made.

The business people who will be using the software need to be more closely involved with those who are building it. They need to clearly spell out what it is they want the software to do. And they need to define those needs early on, and not go back and change or add on to them later.

More companies are beginning to catch on, and they are developing structured programming techniques so that development projects are reviewed more frequently, and adequate time is devoted to testing. As software developer Robert Nagel notes, companies (perhaps after being burned so frequently) are becoming more sophisticated in their staffing, so that they are able to write more detailed and thought-out specifications of exactly what they would like their software to do.

But others disagree. Willbern feels "This is the number one problem facing business information management departments today. I haven't seen any significant improvement after all these years, but they're going to have to learn how to do it."

Even Nagel concedes, "There have been many horror stories, and there will continue to be horror stories."

Chapter Six

COMPUTERS AND UNCLE SAM

To *err is human,* the saying goes, *to really foul things up requires a computer.* Throw the government into the mix, and you have a recipe ripe for disaster.

It turns out businesses are not the only ones that have trouble properly developing their software. Software in government agencies often ends up being error-prone, late, and expensive. Many of the same managerial problems of software development that affect businesses also affect the government. But government agencies are further burdened by bureaucratic obstacles and the sheer scope of their tasks.

The federal government is the world's largest user of software. One out of every four programs is purchased by Uncle Sam. The programs do everything from issue social security checks to launch the space shuttle. So it's probably no surprise that systems so large and complex are rife with software errors. But what is surprising is exactly how much such errors are costing American taxpayers. Taxpayers end up being hurt both directly, as victims of software errors, and also indirectly, by having to pay billions for software development projects that get out of hand.

* * *

In January 1985, businesses in the Northeast began receiving mysterious notices from the Internal Revenue Service stating that they were delinquent on their taxes. Twenty-six thousand eight hundred businesses in Delaware, Pennsylvania, Maryland, and the District of Columbia all received similar letters. The IRS said they had failed to turn in withholdings from their employees' paychecks. The businesses were puzzled, claiming they had paid their taxes on time. Protests to the IRS office in Philadelphia that they were current on their tax liabilities went unheeded. The IRS threatened to begin seizing property if the businesses didn't pay up. At least five businesses, despite their protestations, had their assets frozen. It was another month before the IRS figured out what had happened. Computer tapes containing records of $270 million in tax payments had been misplaced.

Officials at the Philadelphia office finally managed to locate the wayward tapes. The matter was straightened out in due course, and Philadelphia officials figured they had just weathered their quota of crises for the 1985 filing season. They were very wrong. April, after all, was still two months away.

Each year, the IRS is charged with processing nearly two hundred million individual and corporate income tax returns. Such a mountain of paperwork would be impossible to tackle without computerized automation.

At the Brookhaven Service Center in Holtsville, New York, the mailman makes three deliveries a day as filing deadlines approach. Brookhaven handles all returns from New York City, Westchester and Rockland Counties, Long Island, and all of New Jersey. Nearly twenty million returns a year pass through Brookhaven, which is the newest of the IRS's ten regional processing centers.

Resting on a sixty-seven-acre tract of land, the huge five-building complex is home to nearly four thousand people at tax filing time. Many are part-time employees called in to handle the rush as April 15 approaches and millions rush their returns in just under the deadline. Each year Brookhaven handles about seven million individual income tax returns.

As mail arrives on the loading docks, the hefty mail sacks are emptied on to conveyor belts. A computerized sorting machine shuffles the mail at a lightning-fast thirty thousand envelopes per hour. Envelopes are a blur as they whisk through the machine. The Omni-Sorter reads the bar codes printed on the envelopes provided in tax forms, and sorts the mail into separate bins, depending on the type of form that's inside. Form 1040A. 1040EZ. 940. 941. Form 11 series. Correspondence. Form 1040 Non-Business. Fourteen separate categories in all. Uncoded envelopes, referred to as "white mail," are bundled separately.

The Omni-Sorter slits open the envelope as it sends it on to the "Triage" area, known formally as the Extracting and Sorting Unit. At 150 desks, workers remove the contents and begin sorting returns. Mail is split up according to the type of return: single; joint; married, but filing separately. Twenty different categories in all. After the returns are removed, the clerks pass each envelope over a light to make sure it is empty. Before the envelopes are thrown away, another employee also double-checks that they are empty. Returns with no checks are sent to tax examiners who make sure the returns are completely filled out. Those with checks are given to data transcribers who bundle the checks and credit the taxpayers' accounts.

The returns meet up in the data conversion area. Seated at semicircular desks in a room nearly the size of a football field, an army of six hundred data transcribers transfers the information from each return into computers. Each transcriber handles between thirty-five and one hundred returns per hour.

Data is put into blocks, each covering one hundred returns. Twice a day, the computers produce reels of magnetic tape containing information on that day's returns.

Workers then recheck each return for math errors. Incorrect returns are corrected, and the information re-fed into the computers.

Once a week, the computer tapes of the returns are shipped by courier to the National Computer Center in Martinsburg, West Virginia. The information is then added to the taxpayers' master files. The NCC then sends out instructions to the regional centers to send out refunds to the appropriate taxpayers.

It is a daunting task, to say the least.

As early as the late 1960s, the IRS recognized the need to vastly improve its computerized data processing. The IRS's computer system was a mix of old computers from Control Data, General Electric, and Honeywell. As the 1970s wore on, the machines were too slow to deal with the ever-rising mountain of returns. Breakdowns became frequent, and spare parts for the antiquated machines were hard to find.

But an effort to bring in state-of-the-art equipment to handle returns never got off the ground because of a congressional moratorium on the purchase of new computer systems. In the wake of Watergate and the era of illegal wiretaps, Congress feared a centralized computer data system could be used to keep tabs on and harass citizens and businesses. The moratorium was finally lifted in the late 1970s, and the IRS pressed ahead with an ambitious $4 billion plan to completely revamp their computers.

The Systems Modernization Plan would incorporate state-of-the-art high-speed computers. The ten regional return processing centers and the National Computer Center would be linked with advanced telecommunications lines allowing faster access to data, and automated data processing would eliminate most paperwork.

National Semiconductor and Motorola were awarded contracts to build the first stage of the system. Although still a bit crude—all information on a taxpayer's return still had to be keyed into the computer by hand—the system was installed on time and worked fine.

The more ambitious part of the plan started with a $103 million contract awarded to Sperry Univac. New computers would be installed at each of the ten regional processing centers. Eventually, many taxpayers would not even have to file returns. The computers would calculate taxes based on W-2 forms, and then send the taxpayer a printout showing its calculations and setting forth the taxes owed or the refund due. The taxpayer would proof the form, and then either send a check or wait for the refund.

But the IRS plan stumbled coming out of the gate. The contract was awarded to Sperry Univac in the summer of 1981. But the contract specifications were delayed, and not completed until three years later.

By that time, the first computers had already been delivered. Late. The computers refused to work properly. Two of the new Univac 1100/80-series computers were delivered to the IRS regional centers in Memphis and Atlanta. Installers could not get the machines to work. The computers had be removed and hauled back to Sperry.

The tape drives, which stored information on magnetic tapes, all had Sperry labels on them. But IRS officials discovered that a number of them were actually made by Storage Technology, which was in bankruptcy. The machines consistently malfunctioned.

But whatever problems the IRS had in getting its new computers to work were made far worse because of the need to completely redesign the computer software. Programs from one manufacturer's computers often cannot run on another company's machines. The IRS computer programs needed to be rewritten into another computer language. It was a monumental task. The IRS decided to rewrite fifteen hundred separate programs, containing a whopping 3.5 million lines of code, the equivalent of nearly four Manhattan *White Pages* phone books. An army of three hundred programmers, most of whom had little experience with the sophisticated COBOL language, was assigned to the task.

When the Herculean job was finally completed, the new software was tested, first on computers at the IRS's National Computer Center in West Virginia, and then on a new Univac 1100/84 at a regional office in Memphis. The software was then distributed to all IRS regional claims processing centers. IRS officials were horrified to discover the software worked only half as fast as expected. The software had been tested only in its separate component sections. The IRS had failed to test how the programs behaved when they were all operating simultaneously.

All available programmers were immediately reassigned to the project. Programmers raced back to the drawing board and began a crash effort to improve the program.

By November 1984, the IRS leadership had reached a critical turning point. The software rewrite was still being completed. Within weeks, income tax returns would begin flowing into IRS processing centers. IRS officials huddled with time running out on the clock. They elected to go for the long bomb.

The old system had barely survived the crush of returns during the previous filing season. The new system represented the only way to handle the onslaught of paperwork. When it came time to make the fateful decision, the IRS had little choice.

The decision also meant there would be no time for the system to be adequately tested. "There was a tradeoff between getting the computers in as quickly as possible and doing the shakedown runs," said IRS computer chief Thomas Laycock in the *New York Times* (vol. 134, April 13, 1985). The software was going off to the front lines untested, with IRS officials crossing their fingers.

At Brookhaven and Philadelphia and other centers around the country, work quickly began backing up. The new computers worked only about twelve hours out of each day, far less than the eighteen to twenty-four hours IRS officials had been hoping for. Some processors went off-line for days at a time. Because each center had only one central processor, any downtime meant that the processing of returns came to a grinding and complete halt.

The new software programs were excruciatingly slow. In order to correct the problems in the initial software design, the expanded rewritten program took up too much of the computer's capacity, causing the program to run very inefficiently. In addition, it contained a software error that caused employees to have to duplicate some of their work. The computers often crashed. A new software program was supposed to allow workers to pick up from where they had stopped. Instead, the software simply erased what they had been working on, and employees had to start over from scratch. Data transcribers complained they often had to perform work over and over again. The error took weeks to locate.

Problems began piling on top of one another. A serious problem quickly developed in weekend data processing. During the week, legions of data transcribers would enter information from each return into the computers. When Friday rolled around, computer technicians would scramble to process the huge volume of data. Each taxpayer's master data file needed to be updated with the new information. Service centers raced to place the new information on computer tapes so they could be shipped each Monday to the National Computer Center files in West Virginia.

Each center had only about sixty hours each weekend to complete the job before workers would come in Monday morning to start work on new returns. The computers were so unreliable the work often spilled over into the work week. "It was taking them considerably longer than the forty-eight-to sixty-hour period to get things done," James Watts, Associate Director of the General Accounting Office, recalls. "The computers weren't available to the workers Monday morning, so people couldn't process more tax returns."

"Before we knew it," Bob Hughes, the director of Brookhaven told the *New York Times* (vol. 134, April 15, 1985), "we were seven days behind." Other regional processing centers had similar problems, some of the worst backlogs being experienced in Fresno and Austin. Millions of tax returns were getting "hung up" within the computers. Clerks in the error resolution department needed access to computer files to correct math errors that people had made on their returns. But because of a software error in the computer programming, half the returns could not be located on the computer.

As W. C. Fields might have said, things were naturally far worse in Philadelphia. With three weeks to go in the tax filing season, the Philadelphia center had completed work on fewer than half of its returns. Faced with the prospect of a crushing flood of returns that would come cascading in during the final weeks, when nearly half of all taxpayers file their returns, managers and workers panicked.

The Philadelphia Inquirer reported that frustrated and overworked IRS employees had destroyed some returns. Commissioner Roscoe Egger vehemently denied that at an April 10 news conference. "Sheer, utter nonsense," he called it. "We would know it from our accounting system if returns were destroyed."

It took the General Accounting Office several months to piece together the true story. On April 26, a Philadelphia employee found unprocessed returns in a trash barrel on the center's loading dock. A hundred and nine envelopes were inside, containing thirty-six individual tax returns, and ninety-four checks, some as large as $66,000, totalling $333,440. The envelopes had been too large to fit through the Omni-Sorter, and frustrated employees had simply tossed them out. Four days later, an IRS auditor found three envelopes containing $2,500 in a different trash barrel. A mail clerk was fired for what IRS supervisors called "continually disposing of taxpayers' forms and checks in the wastepaper basket."

GAO investigators found that as the Philadelphia Center became backed up with a growing mountain of unprocessed returns, workers had been pushed to the limit. Mail processing employees had been working up to nineteen consecutive days without a day off. Their two supervisors worked consecutive eighty-hour weeks. They were under pressure to perform, and had handled the workload by throwing some of it away.

It wasn't the first time, either. A year earlier, ninety-two returns had been thrown away in wastebaskets in the women's restrooms in the Philadelphia Center.

The GAO was not able to confirm reports that thousands of returns had ended up being intentionally shredded. But over a thousand taxpayers did have to refile their returns because the IRS had no record of ever receiving them. Whether the returns were destroyed by frustrated employees or simply lost in the computer shuffle may never be known.

Around the country, the IRS was stuffed with a backlog of twenty-four million returns with one week to go. Of sixty-one million returns that had arrived, only some thirty-seven million had been processed. Forty million more returns would arrive in the next week. Refund checks were frequently going out a month late. In the IRS processing center in Andover, Massachusetts, a worker hung up a sign: "THE DIFFERENCE BETWEEN THIS PLACE AND THE TITANIC IS THE TITANIC HAD A BAND."

Through it all, the IRS maintained a stiff upper lip. Even as the calendar edged towards April 15, the filing deadline, Commissioner Roscoe Egger held a news conference to give his assurances that nearly all refunds would be mailed on time. He had a powerful incentive motivating him. By law, if any refunds are not mailed by June 1, the IRS must pay thirteen percent interest.

Egger lost his bet in a big way. Despite a late surge by the IRS in processing returns, the agency finished the race well behind. More than $42 million had to be paid out in interest on late refunds.

The delays created a domino effect that rippled through the entire economy. The government noted that consumer spending dropped noticeably in March and April, in large part because nearly $7 billion in tax refunds was late arriving in consumers' hands. A drop in the nation's money supply brought on by the absence of refund money forced the Federal Reserve to adjust its economic forecasts for the coming year.

Computer experts berated the IRS for trying to install both new hardware and software at the same time. They also criticized the IRS for not running its old system alongside the new one, as a backup. IRS officials said the cost of running both systems would have been prohibitive. As IRS computer chief Thomas Laycock said to the *New York Times* (vol. 134, April 13, 1985), "We gambled and lost."

The end of the tax season did not bring an end to the IRS's computer problems. The tangle of backlogged returns and lost records took months to straighten out. Over the ensuing winter, the IRS lost another computer tape, holding information on $185 million in income tax withholdings. This time, the IRS was able quickly to find the stray tape, which had been waylaid on its journey from a Southwest processing center to the National Computer Center in West Virginia.

By the time the next February rolled around, the IRS still faced a backlog of 2,154 unprocessed returns and 1.1 million unanswered letters asking for tax help, left over from the previous year. Incredibly, the IRS was still struggling with 5,004 returns still left unprocessed from the 1984 tax season, two years earlier.

In the face of it all, Commissioner Egger brushed aside tales of discarded returns and misplaced records. "There were a few isolated instances of that," he said to *U.S. News and World Report* (vol. 104, November 18, 1985). "Allegations of destruction were much more plentiful than the actual occurrences, and the impact on taxpayers was very limited."

Egger was looking beyond the short-term problems towards the future. No one questions the IRS's need to vastly upgrade its computers. Its old system was based on designs from the fifties and equipment from the sixties. After extensive reworking, the Sperry processing computers and the new COBOL software worked smoothly in 1986.

Egger felt the new system laid the groundwork for a system of electronic tax filing that could eliminate most of the paperwork for many taxpayers. "Someday," Egger confidently predicted in *Business Week* (no. 2892, April 29, 1985), "you'll be able to sit at your home computer and file from your living room." Egger foresaw a day when more than half of the nation's individual taxpayers would file electronically. Such a system could be a boon to taxpayers, who every year spend a staggering total of ninety-seven million hours slaving to fill out their tax forms. The IRS estimates it takes the average taxpayer seven hours and forty-four minutes to complete the standard 1040A form, and a still-daunting two hours to fill out the simpler 1040EZ.

There is also the lure of savings to taxpayers. Electronic filing holds the promise of reducing some of the incredible $2 billion we pay each year to professional tax preparers.

Electronic filing is here now, sort of. In 1990, the IRS for the first time made filing by computer available in all fifty states. The system is not yet in the form the IRS envisions. Taxpayers must still work out their calculations on paper, and then physically take them to one of eighteen thousand tax preparers who are specially equipped to file returns by computer.

The IRS can then electronically direct-deposit the ensuing refund into the taxpayer's bank account. The fee for the service is from $25 on up. Electronic filing could someday be the salvation of the IRS, which each year hires thirty thousand temporary employees to help process returns. Processing an electronic return costs only a fraction of its paper counterpart.

But even though its use doubles with each passing year, fewer than two percent of all taxpayers use the service. Many tax experts feel even the new simplified tax codes are too complicated to make widespread electronic filing feasible. Thomas Block, president of H&R Block (*Business Week*, no. 2892, April 29, 1985) said, "We'll probably never see the level of reform needed to make paperless filing a common method of filing."

And even several years after the horrors of the IRS's version of *The Philadelphia Story*, its computer system is still disturbingly susceptible to human and computer error. In the 1990 filing season, an employee in Cincinnati who delayed running one day's worth of electronic returns delayed returns for fifty-three thousand taxpayers for two weeks. During the same month, a faulty computer tape in the center in Ogden, Utah delayed 171,000 returns. IRS spokeswoman Ellen Murphy said about a dozen computer tapes around the country contained similar faults.

Money magazine (vol. 19, April 1990) estimates taxpayers pay $7 billion annually to the IRS that they do not owe. Indeed, a 1988 GAO study found that nearly half of the IRS's responses to taxpayers' letters were inaccurate. IRS Asst. Commissioner for Returns Processing Charles Peoples told the magazine, "We don't have good methods for detecting and correcting errors." Even after massive efforts to improve the training of IRS employees, there is still an error rate of at least twenty percent in IRS written responses in cases where the IRS is demanding extra taxes, interest, or penalties. IRS chief Fred Goldberg admits many taxpayers end up paying more than they owe, although he says he hopes the amount is far less than the magazine's figure.

Meanwhile, the General Accounting Office reports the IRS is owed more than $66 billion. Many taxes have been assessed, but are unpaid. More than thirteen million individuals, couples, and businesses owe taxes that have gone uncollected. No one, Goldberg admits, knows the precise amount because the figure likely includes duplicate and erroneous billings. And the gap continues to grow, more than doubling over the last five years. As Assoc. GAO Director Paul Posner puts it, "Each year, IRS falls further and further behind in its efforts to reduce the balance."

IRS officials say these incidents show the need to continue their computer modernization. "This points up once again," said IRS spokesman Frank Keith, "that our computer system is antiquated."

But no matter how advanced the IRS's computer system ever gets, there is always room for new errors to be invented.

Mary Frances Stinson found this out in March 1990. The Dorchester, Massachusetts native received a form letter demanding she pay the amount she was short on her taxes three years ago. Ten cents. The computer then made matters worse by tacking on a twenty-four dollar late payment penalty and a thirty-one dollar interest charge.

The computer didn't even get its own facts straight. In the letter it noted that, by law, no penalty can exceed twenty-five percent of the late payment, which in this case would mean two cents.

An IRS spokesman sighed, noting that even the postage needed to obtain the loose change would be more than the overdue taxes.

IT's not surprising to learn the IRS is not the only government agency having problems with its large software systems. As head of the General Accounting Office, the investigative arm of Congress, U.S. Comptroller General Charles Bowsher regularly looks into how effectively the government is spending its money on computerization. He notes with some discouragement, "Some of the government's major systems projects have had costs escalate by hundreds of millions of dollars and schedules slip by years. In many cases, the new systems do not work as planned, and waste millions of dollars."

The government could not function long without computers. They are a vital part of our defense. In addition, the federal government relies on computer software to deliver social security checks to the elderly, provide benefits to disabled veterans, direct aid to needy families, launch space shuttles, coordinate air traffic control, and track criminals. It is impossible to imagine government trying to conduct its affairs without the use of computers.

But Uncle Sam's computers are riddled with software problems.

- One hundred twenty-five million Americans pay social security taxes. Thirty-eight million Americans receive social security benefits. A major modernization program was started in the Social Security Administration in 1982 in an attempt to efficiently handle that crushing workload. After spending more than $500 million on new computers and software, the SSA is worse off than ever. Its computer systems are in what the House Government Operations Committee called ''a dangerous state of disarray.''

  It takes twice as long to complete paperwork under the new computerized system. The system is barely able to issue checks on time. One out of every six beneficiaries is either underpaid or overpaid in benefits. Errors in benefits and postings of contributions go unnoticed for years.

  Much of the disarray is due to a failure to properly design the software for the system. One top Social Security Administration official said a lack of understanding of the type of data used by the SSA resulted in ''a lot of time spent running around in circles'' trying to develop software requirements.

  As a result, the software was often error-plagued and outdated. Kenneth Blaylock, president of the American Federation of Government Employees, glumly stated, ''SSA's modernization program is a shambles. The heart of the modernization, software development, is years behind schedule. The agency will still be using its admittedly inadequate software for a lengthy period well into the 1990s.'' Blaylock further noted, ''Service to the public is suffering. Millions are being underpaid.''

In 1988, the House Government Operations Committee agreed, saying, "It is unlikely SSA Commissioner Hardy and other senior officials of SSA will be able to salvage the Systems Modernization Program. More than $500 million has been spent . . . and there is very little to be shown for it."

• The federal government is planning to spend up to $25 billion to upgrade its telecommunications network. The program, known as FTS 2000, is to serve the long distance needs of government agencies well into the twenty-first century. But the General Services Administration failed to adequately study alternative systems. As a result, the General Accounting Office concluded there was no way of knowing if FTS 2000 was the most economical or even technically the best system. The GAO is recommending the government make no long-term commitments to go ahead with FTS 2000.

• The Veterans Administration spent $200 million to computerize its 169 medical centers around the country. The software for the computers was not adequately tested and was deemed to be susceptible to undetected errors. Several revisions had to be made to the software because it had been prematurely been put into operation. GAO concluded the most cost-effective system may not have been selected.

And the problem of cost overruns in government computer systems is almost exclusively one of software, rather than the computers themselves. As the House Committee on Science, Space, and Technology notes, in nearly all government agencies, "today, what has become apparent is that system costs are driven not by hardware, but by software."

Why does Uncle Sam have such a difficult time developing its software?

Part of the difficulty stems from the sheer size of the federal government. Government agencies, and their computers, dwarf even their largest counterparts in the Fortune 500. Likewise, the software needed to run the computers is called on to perform gargantuan tasks without parallel in the private sector. The Veterans Administration is the largest provider of health care in the United States. The Social Security Administration has to keep track of benefits for thirty-eight million Americans who depend on their social security checks. The Internal Revenue Service has to keep records for seven years on two hundred million taxpayers.

Faced with such immense tasks, federal software projects typically exceed in cost, size, and complexity any comparable systems in the private sector. As GAO notes, "many federal systems are being designed as one-of-a-kind in their size and complexity." Such huge systems not only demand highly complicated software, but leave the government with few real-world models for trouble-shooting.

Consider the Internal Revenue Service. Its software has proven the nemesis of many a beleaguered taxpayer, such as Mary Frances Stinson and her overdue dime. But the IRS serves as a good example of both the compelling need to overhaul federal computers and the difficulty in doing so. The problem is often one of sheer size, and the immensity of the task of replacing these silicon chip behemoths and their million lines of code software.

Let's go back to Martinsburg, West Virginia.

"This is the one place where it all comes together and where it all goes back out," says Gerald Rabe as he surveys the vast complex that is the National Computer Center. The gargantuan one-story building sits less than a two-hour drive from Washington, D.C. amidst the peach orchards in the Shenandoah Valley of West Virginia. The system is aptly referred to as "The Martinsburg Monster." Row upon row of Sperry Univac computers and tape drives stretch throughout the complex. The machines are clearly not state-of-the-art. Many of them are no longer manufactured, and spare parts are difficult to find. The old tape drives are the type one would have commonly seen used in businesses in the 1960s.

Still, the task the NCC is called upon to perform dwarfs that of nearly any other computer system in the world. As NCC Director Rabe explained in an interview with the Associated Press, ''I would say the computer system we use here is probably second to none in the amount of data processed.'' That is scarcely an exaggeration. The magnetic computer tapes here hold the tax records for the previous five years of 140 million individuals and 50 million businesses.

The statistics of NCC are staggering. Nearly two hundred million tax returns are processed every year. Nearly $1 trillion in tax revenue is collected. Sixty-five thousand computer tapes are delivered to the center from the IRS's ten regional offices every year. The NCC's central master file of eighty-eight hundred reels of magnetic tape would stretch around the world more than five times. Walking into the basement, one can stroll past row after row of thousands of computer tapes in storage.

Merely the fact that there are computer tapes points out the antiquated condition of the IRS's computer system. Instead of having high-speed telecommunications links between the National Computer Center and its ten regional service centers, information on tax returns must be transferred to magnetic computer tapes. The tapes are then taken to the airport, loaded onto airplanes, flown to West Virginia, and then trucked to the NCC. ''It's a twenty-five-year-old design,'' complains GAO's Assoc. Director of Information and Technology Division James Watts. ''It's highly inefficient, it's extraordinarily costly. It really needs to be redone from the ground up.''

National Computer Center director Gerald Rabe says he foresees the Martinsburg Monster eventually transformed into a streamlined state-of-the-art system. Instead of hauling thousands of computer tapes between service centers, information will be beamed from NCC computers to the ten regional offices. New software will offer impressive instant access to taxpayer files. ''We want to get to the point,'' Rabe says, ''where banks and gas and electric companies are. If you call us with a question, we want to be able to call up your entire account on the computer screen and answer your questions immediately.''

Asst. IRS Commissioner Phil Wilcox concedes what most government officials already know. "We know the system's old. We know it's outdated and needs to be replaced. Our Systems Modernization Plan is designed to do exactly that." The General Accounting Office called the IRS's lack of modern computer tax processing systems the "IRS's most pressing overall need."

But others are less convinced the IRS can be entrusted to develop an $8 billion computer system. Even in the wake of the 1985 debacle, the IRS continued to have problems developing its new equipment. Severe software problems were found in new computer processors installed in November 1986. The computers failed an acceptance test; they could not handle the volume of data required. A contract for five thousand computer terminals had to be relet, delaying the project at least a year.

The IRS's hoped-for salvation, the Systems Modernization Plan, begun in 1981, is running over budget, is fraught with delays, and still shows few signs it will accomplish its objectives. The nearly $8 billion price tag is already double the original estimate and some critics charge that the final bill could go as high as $11 billion. The project's completion date has been pushed back to at least 1998.

In November 1988, the GAO warned that the IRS's project "is expected to take about ten more years to complete and cost several billion dollars. [And] those estimates could change dramatically, because IRS had not, as of August 1988, identified alternative designs, made cost-benefit analyses, and selected a final design."

House Subcommittee on Oversight Chairman J. J. Pickle (D-Tex.) shares those fears. "Over the next six or ten years," he told a subcommittee budget hearing in 1987, "the IRS is faced with restructuring its entire computer systems so as to better use its resources and serve the public. Yet, recent events raise real questions about whether the IRS has the resources, the discipline, and the support to meet its current workload demands, while also planning for and acquiring future state-of-the-art computer systems."

And given the mixed history of such large government software projects, some officials question whether the system will ever evolve into what its developers had envisioned. As GAO's Watts says, "The IRS has really struggled in this area, and has an awfully long way to go before there's going to be any major changes." The IRS will have little choice but to continue using its outdated computer and telecommunications systems, which are in danger of reaching capacity.

Much of the difficulty in federal software development stems from a lack of coherent vision of what federal agencies want their software to do. Long-range strategies are often lacking. Ninety percent of senior IRS executives surveyed felt the agency did not have a clear focus on how it wanted to manage its information resources. The IRS's projections for its workload, used for setting the size of its computers, were found by GAO to be "outdated and could not be relied upon."

Part of the problem is the constant turnover of high-level government officials as administrations come and go. Watts noted, "several leadership changes have occurred both within the IRS and the Treasury Department [over the past several years]. As a result, the approaches to implementing a new system have varied in accordance with the preferences of leadership."

"That's a major problem, not only with IRS but with other agencies," Watts says. There often is a lack of one consistent leader who can shepherd the project all the way from inception to completion. "When you have turnover of people in key positions, that has a tendency to redirect things and slow things down as you go through a project. These projects are things that take several years to do."

And it's more than simply a problem of turnover within the agency itself. Projects at the Internal Revenue Service must be cleared through the Treasury Department. Similarly, plans for large computers at the Federal Aviation Administration have to work their way through the hierarchy of the Department of Transportation. Food and Drug Administration computer development projects have to be cleared through Health and Human Services, and so on. Any change in leadership at those departments can spell delays and detours for software projects in their respective underlying agencies.

And the turnover does not have to be at the highest levels of management to spell disaster for software modernization. Managers are simply not able to hire and hold on to the qualified technical personnel they need to form projects and see them through to completion. In a 1987 survey, nearly forty percent of federal managers complained that their ability to hire needed staff had worsened in the last five years. They noted the problem was particularly acute in computer science. The GAO reported with alarm, ''The shortage of trained personnel is already apparent and is projected to become severe by the year 2000.''

The problem, GAO noted, was especially critical, because several federal agencies are in the process of converting their systems to more technically complex systems that will require a high level of expertise.

But at a time when the federal government needs them the most, Uncle Sam cannot pay talented personnel enough to compete with the private sector. Government salaries for computer specialists are far below those offered elsewhere. The President's Commission on Compensation of Career Federal Executives found that private industry pays senior executives as much as sixty-five percent more than the government. The Merit Systems Protection Board found that computer specialists in the government earned between twenty-two and thirty-two percent less than those in private corporations.

Officials at NASA find that sometimes they can recruit qualified people only because the applicants are fascinated by the space program and willing to accept the lower pay. Still, NASA often has to bend the rules to bring technicians on board. Linda Ragsdale, NASA's supervisor for staffing, reports NASA tries to find ways to classify incoming software specialists as engineers, qualifying them for higher pay levels.

The unflattering image of public service also aggravates the turnover problem. A 1987 GAO survey found that negative public perceptions of government work were the greatest source of job dissatisfaction among senior executives.

A lack of stability among senior executives and inability to hire qualified technicians contributes to the problem of federal agencies not being able to figure out what they want in their software systems. Further, GAO notes, agencies often fail to check with the people who will be using the system to see what they want. Not surprisingly, the result, according to the 1987 survey, is that "agencies frequently do not adequately define their requirements, and therefore implement systems that do not meet users' needs."

Because of this lack of coordination, managers tend to go back and ask for changes while the software is being developed. GAO's James Watts says, "This happens across the board in any agency. The people upfront have a difficult time really laying out their requirements of what they want the system to do. And then after a project is underway, people say, 'Well gee, I've got this other thing I'd like it to do.' So they start adding on requirements after the project has already moved down the pike. Trying to add them in a lot of instances creates delays, redesign problems, and cost increases." One Treasury Department system mushroomed in cost from $8.5 million to a staggering $120.5 million. By the time it was cancelled, it was also eighteen months late. Treasury officials blamed a lack of coordination by senior officials, and a failure to consult with the system's intended users.

Obscure and confusing government regulations and procurement procedures hardly help. In 1989, Amtrak was having problems coping with projected traffic increases in the heavily travelled Northeast Corridor between Washington and Boston. Amtrak was in the process of developing a computer-controlled switching system that would greatly increase traffic capacity.

But the contractor, Chrysler, was having difficulty developing the software. As one government official wryly noted, "Chrysler is not typically the first company that comes to mind when discussing software questions." According to project manager James Early, chief engineer for the Northeast Corridor, Chrysler was not even the recommended choice for the contract. But the Federal Railroad Administration had ordered Amtrak to accept the bid. The reason: Chrysler had come in with the lowest bid.

In 1986, the IRS found major errors in its price evaluations of computer contract proposals. It also overpaid on a data processing contract by nearly $1 million. Officials said they simply had no idea how much they had been paying on the contracts.

Therefore, it's not surprising that the government's few attempts to institute standards for software development have generally been failures. In February 1988, the Defense Department wrote a standard known as STD-2167A. It attempted to clearly define how software was to be ordered. The government hoped STD-2167A would enable agencies to better define what they wanted, and give contractors a clearer blueprint of what they were expected to develop.

But STD-2167A proved to be so unwieldy that the government often had to turn to the contractors, and ask *them* how the agencies were supposed to apply the standard to their proposals.

As a recent report from the House Committee on Science, Space, and Technology stated, "Software cannot be properly developed using the welter of regulations presently in force. The procurement system as presently structured does not take into account the special needs of computer software systems and can compromise effective software development from the start."

Good software development requires that money be spent upfront. Considerable time and effort has to be spent painstakingly going over the design of a program to make sure it will work effectively, be adaptable to change, and do what everyone wants it to do. More money has to be spent in the planning stages. This often makes projects *appear* to be more costly and slow to develop early on.

Such thinking, however, works contrary to the budgeting process. In order to secure initial funding, project costs are often purposely underestimated. Planning costs are usually the first to go. The savings will often come back to haunt a project when it undergoes extensive redesign later on, and costs and delays climb.

The end result is summed up by House investigators James Paul and Gregory Simon: "The government does not capture the full benefits of its investment in computer technology. Government managers find that the software they buy or develop does not achieve the capabilities contracted for. It is not delivered at the time specified, and the cost is significantly greater than anticipated. Many recent examples of cost overruns and degraded capability cited as examples of government waste, fraud, and abuse can be related to problems in the development of computer software."

Sen. Albert Gore (D-Tenn.) has been involved in the government's investigation of software procurement practices for some time. He says, "The federal government is the country's largest consumer of software. So clearly we would save taxpayers' money with better software that is easier to use, easier to write, and less prone to errors."

As in most instances of computerization and software development, this is easier said than done.

Chapter Seven

UP IN THE AIR

GOVERNMENT computers do far more than simply take our money and defend us. We also heavily rely on them for our personal safety. By and large, they do a commendable job. But perhaps nowhere is reliable software more important than in air traffic control. Millions of lives depend on it every day.

The nation's air traffic control system could not function without computers. Automation is virtually the only way to handle the incredible flow of more than 450 million passengers that crisscross the nation's skies every year. Computers help air traffic controllers keep planes on track and on time. Altitude, positions, flight paths, and weather conditions are all tracked by computer. But the complexity of the systems makes them vulnerable to software problems. And the results can make air traffic controllers nervous.

Go Sooners! Whooeeeee!
   Go get 'em, Longhorns! Hook 'em horns!
Once every year, it seems, the massive Dallas-Ft. Worth International Airport turns into one giant tailgate party. Thousands of Sooner and Longhorn fans, boosters, and alumni stream through

the terminal corridors, making their annual pilgrimage to the year's biggest party, the Texas-Oklahoma college football game. The loud, boisterous fans, many sporting ten-gallon hats, colorful pennants, and horns, make their allegiances noisily apparent to friends and strangers alike as they migrate through the seven terminals. Since the dawn's early light they have been arriving from all over the country, snaking their way to the people movers that would take them around the vast airport grounds on their way to the final destination, the Cotton Bowl stadium.

The airport, referred to by locals simply by the code stamped on baggage tags, DFW, was up to the task. Like everything else in the state, the airport is Texas-size. The sprawling complex is the largest airport in the United States, covering more land than Manhattan. It ranks also as the nation's second-buslest airport, behind only Chicago's O'Hare. At rush hour, more than three hundred flights move in and out every hour.

But the wonder was, Milt Scherotter reflected, that it was operating at all this day. For there were Texas-size problems in air traffic control. Milt was working the day shift at the Tracon facility located at the base of the DFW tower. Terminal Radar Control, or Tracon, guides nearly every plane arriving or departing the vast Dallas-Ft. Worth metroplex area. Milt was handling the North Dallas sector, handling planes arriving at Love Field near the downtown. Love Field, formerly Dallas-Ft. Worth's main airport, hadn't seen much action since the big boys moved their operations to DFW when it opened in 1976. Oh sure, an occasional Southwest Airlines flight would come in from Houston or El Paso, but Love was now mostly home to light airplanes and fleets of corporate jets belonging to what remained to Texas's battered oil and banking giants.

But today, Milt knew, was going to be different. Since he'd arrived at 6:30 that morning, he had been mentally bracing himself, for what he knew was coming. He had worked a Texas-OU game days before.

Hours earlier, at the crack of dawn, hundreds of pilots had climbed into their planes and taken off from places like Oklahoma City,

Norman, Austin, Houston, and San Antonio. Soon they would begin converging on Dallas-Ft. Worth from all points of the compass like an invading army. Everything from single-engine Cessnas to speedy corporate jets.

Most would be heading for Love Field because of its closeness to the Cotton Bowl stadium. But some would scatter to smaller airports. Meacham. Red Bird. Mesquite. Addison. Grand Prairie.

So far, the Automated Radar Terminal System, or ARTS IIIA, was handling things just fine. On the radar screen in front of him, Milt watched dozens of radar blips inch across the scope as planes crisscrossed the airspace over the metroplex. Radar had come a long way since controllers had to use strips of paper to keep track of flight information. The ARTS IIIA software generated tiny alphanumeric characters that appeared alongside each blip, or target. The display showed airline, flight number, altitude, and airspeed. As the target moved, the information crept alongside of it. Within a small box at the bottom of the screen, the computer displayed the time, current weather information, and the status of the runways at the various airports. The information was vital for controllers to keep track of planes, and it greatly reduced their workload.

Milt knew how the drill worked but that didn't make it any easier to cope with today's extra work. Ever since about 8:00 A.M., the traffic had been picking up steadily. So far so good, but traffic was starting to get heavier.

The veteran controller just hoped the computer could handle all this traffic. To make matters even worse, the biggest convergence of general aviation traffic coming in for the Texas-Oklahoma University game would coincide with the morning rush hour of flights going in and out of DFW. For the airlines with the biggest operations at DFW, such as American, Braniff, and Southwest, this would be the time when all their flights would be arriving and departing. As one controller had put it, "It seems everybody in the world wants to fly from Houston to Dallas-Ft. Worth and arrive at 9:30 in the morning." The nation's second-most-crowded airspace was about to get even worse. Soon, more than one hundred planes would be filling the skies near Dallas-Fort Worth.

After 9:00 A.M., Milt thought, *this place will be hell.*

Milt looked around the dimly lit, windowless Tracon room. The green phosphorescent glow of the banks of radar screens outlined the fifteen other controllers on duty, about normal for a Texas-Oklahoma weekend. Milt wondered why his superiors hadn't bothered to bring in more people. They could sure use them. Especially if the computers went down again.

As most of the controllers were well aware, the Dallas-Ft. Worth Tracon Center had been straining for years under an antiquated computer system. The old Univac 8300 was strictly 1960s-vintage. The machine was woefully short of the memory capacity needed to handle traffic in the nation's second-busiest airspace. With alarming frequency, the computer had overloaded and caused all information to disappear from the computer screens. It usually never lasted longer than thirty seconds.

FAA officials had worried it could happen again. Just one year earlier, the extra workload of the Texas-OU weekend had caused the computer to crash. In the words of one FAA spokesperson, "It just sort of maxed out." Harried controllers had to track flights using handwritten notes for about ten minutes. The FAA was determined it would not happen again.

A computer specialist had been brought in. He had written a software "patch," a revision to the computer programming that would cut out some non-essential radar services and reduce the workload on the overstressed computer processors. When the morning rush hour began, supervisors watched as the traffic pushed, and then exceeded the red line.

The processors were straining at 105 percent of their rated capacity. Only designed reserve capacity was keeping the system afloat. At about 9:45 A.M., supervisors gave the go-ahead to insert the patch and try and relieve some of the pressure on the processors.

Deep in the heart of the computer center, the specialist gave the necessary commands. The trouble was, he entered them into the computer from a terminal that was not authorized to give such commands. The computer looked at the instructions, saw where they were coming from, and quickly became confused. The confusion pushed the over-strained computer over the edge. It maxed out. The computer had simply had enough.

A noisy alarm horn cut through the charter of the Tracon room. All of the controllers froze. They had heard it many times before. The automated computer system was about to crash.

Oh, shit! Here we go again!

Expletives filled the air as controllers braced themselves. The sound of people frantically shuffling papers filled the room. The controllers all knew what was about to happen.

Milt started making mental notes about the planes that were flying in his sector. He would need it.

Three seconds later, all computer information disappeared from radar screens in the Tracon room.

Better start prayin', boys. We're gonna need it.

When the Automated Radar Terminal System came back to life a few seconds later, it was a mess. The software had in essence suffered a nervous breakdown. It was now performing unpredictably. Some flight information was missing. Some data was distorted into gibberish. Some information appeared normal, but was "frozen" and refused to move alongside the radar image it was assigned to.

Don't think. Just act.

Milt quickly keyed his microphone. "Attention all aircraft on this frequency," he broadcast. "We have just lost our automated flight information." He would now have to guess which airplane was which, he could no longer tell them apart. There was no time to even think, let alone panic. Milt had to quickly reconstruct the information the computer had just lost.

"Do not listen for radar advisories. Keep your eyes outside your aircraft. We can no longer guarantee traffic separation." For the next few minutes, the radar would be virtually useless. Radioing a command to the wrong airplane could result in a mid-air collision. Until the controllers could figure out which plane was which on their scopes, the pilots would be on their own.

Afterwards, the FAA would say that controllers had reverted to backup systems upon the failure of the ARTS IIIA. Milt reached for his "FAA backup system." A pad and pencil. He began furiously scribbling down as many aircraft call signs as he could remember.

Cessna Two-Five Bravo . . . Southwest Two-One-Two . . . Cherokee Four-Five Lima . . .

Milt reached across his console and picked up a phone that was a direct land link to the Love Field tower. "This is DFW Tracon. We've just lost our automated computers. I need you to hold all departures. You got that? I mean ALL departures. Nothing goes out!" The last thing he needed was additional flights to worry about. Similar calls were going out from other controllers to DFW tower and to Ft. Worth Center, which handled nearby en route traffic. The Center immediately began "spinning" Dallas-Ft. Worth-bound planes, hastily giving new courses that would keep them out of the local airspace. The controllers had their hands full enough trying to figure out what they already had.

Start scramblin', Milt. You ain't got any time to dink around.

Some of Milt's high-speed jet traffic was moving at nearly three hundred miles an hour. If it took him a minute or two to sort everybody out, some of the planes could move ten miles.

"I'm going to start reading off call signs," he radioed. "You give me a holler if I forget you." Milt began listing airplanes from the notes he'd hastily scribbled.

One by one, the airplanes in his sector called back on the radio. He had only missed one, and that wayward plane quickly reported in. *Geez*, Milt thought, *so far so good. Now what do we do?*

Instinct told Milt to start separating the traffic by altitude. If he assigned each plane a different altitude, it would reduce the danger of mid-air collision while Milt tried to regain his bearings. Across the room, controllers handling arriving jetliners into DFW didn't have that luxury. Arriving flights came in at only two altitudes, eleven thousand feet and twelve thousand feet, depending on whether they were east-bound or west-bound. DFW's five arrival corridors were sandwiched vertically in between paths used by planes departing DFW, which go out at ten thousand feet and thirteen thousand feet. Assigning new altitudes could send flights directly into the paths of departing jets.

Milt had suffered through computer failures before. Hell, there'd been six of these in just the last two months. Usually they lasted only about thirty seconds, max. Milt knew this one was different.

Big time.

"It looked like somebody had blown up my computer screen," Milt said. Numbers and letters were scattered over the screen like debris from a grenade explosion. A clock readout that usually rested in the lower center of the screen now sat on the left edge of the display. Barometric pressure reading information had moved to the right margin. And listings of active runways had been "blown" from the bottom all the way to the top of the screen. Milt now had reason to doubt the accuracy of all of the information that was displayed on the screen.

The software was now in an operating mode engineers refer to as being unstable, meaning it was functioning erratically. Milt could no longer tell which information was accurate and which was false. Some of the data on his screen was obviously junk; it had been converted into gibberish. But other data still appeared to be correct. Milt started making mental notes of which flight information was right and which he would have to ignore.

He glanced up at the controls of his computer display. If he turned down the luminance intensity on the scope, he could lose all the information being provided by the ARTS software. *Naah, some of the stuff is still good. Maybe it can help me out.* Using the radio, Milt confirmed as much of the information as he could.

On the other side of the room, at the console of a controller handling DFW Approach Control, the computer had mixed up the flight numbers of two American Airlines jets. The software switched the numbers, and had the wrong information attached to each blip. The controller, unaware of the error, commanded the wrong plane to begin its descent to approach for landing.

A controller for Departure Control watched in horror as the plane descended into airspace usually packed with flights departing DFW. They were lucky. The particular departure corridor was empty at the time. Minutes earlier, it had been jammed with a long string of departures that would have directly crossed the path of the errant jet.

Milt was starting to get a handle on sorting things out. By now he had prioritized his list of planes according to which was closest to Love Field. The computer screen flickered every few minutes. The technicians were obviously trying to reset the system and restore order. But nothing changed. Each time the screen flashed, the problem remained.

He could feel sweat starting to bead on his forehead. *This ain't too bad actually, handling about six planes.* Still, juggling the planes in his sector required some tricky mental gymnastics.

Let's see now. That Southwest 737's ready to be turned on to final approach. The Learjet can be slotted in right behind him. Keep an eye on that Cessna. He'll be okay for a couple of minutes, but then he needs to be turned left, and then left again.

Damn it to hell. Why does this always happen at the busiest time of day? The computers never go down when it's quiet!

At Milt's previous posting at Los Angeles Tracon, the computers had spit out paper strips that contained flight information for each plane. The strips made a handy backup for just this type of situation. At Dallas-Ft. Worth, which was more highly automated, the strips normally would have just gotten in the way. Except now.

One lucky thing, Milt thought. Just before the computers went down, he had asked for help in dealing with the rising flood of traffic in his sector. A second controller had slid into the position just to Milt's right. Milt had handed over about a half-dozen planes, leaving him just six or seven to deal with. They had completed the switch just ninety seconds before the computers crashed. Together, the two men quickly reconstructed the flow of traffic in their particular sector.

He was glad he'd handed off the other six. The six he had right now were about all he could comfortably handle. *That would have been too much. If I hadn't . . . Naah, I don't even want to think about that.*

It was now twenty minutes since the crisis had started. Controllers at Tracon were little by little gaining the upper hand. Arriving flights were carefully guided into landings at DFW and Love Field. By cutting off new departures, as well as new arrivals from other airports, the traffic began to thin out to more manageable levels.

The floor supervisor was going around the room telling controllers to prepare themselves. Technicians were going to have to "cold boot" the system. The computers were to be completely shut down and then restarted. All data in the computers would be lost for two minutes, and the system would then be restarted from scratch.

What the hell. The whole system's already backed up. May as well flush the whole damn toilet.

The cold boot worked. All ARTS information vanished from the screens for two minutes. But when the system came back up, it had been restored to its normal operating condition. Still, it would take hours before controllers and airline dispatchers could untangle the jumble of hundreds of flights that had been delayed and diverted.

Instead of feeling relief, Milt Scherotter was angry. "I was pissed," he recalls. As soon as the traffic had been reverted to a near-normal order, Milt demanded to go off position for a short break. Using an outside phone line, he called Ed Mullen, the regional representative for the controllers' union. Ed wasn't in, but Milt left a message on Ed's pager for him to call Milt immediately. Mullen called back quickly, and Milt related what had happened. The two men agreed that controllers should file reports to get the incident officially recorded. Before the week was out, all sixteen controllers on duty filed safety incident reports with federal authorities.

Milt feels the margin of safety in the skies over Dallas-Ft. Worth that day had been drastically reduced. "There's no question in my mind conditions were not as safe as they normally were," he says. "The fact planes were separated at all was basically left to chance. It was a scary-as-hell feeling." Only good weather and clear visibility, and some good luck had prevented a disaster.

Six days after the Texas-OU incident, controllers at DFW held a news conference, publicly demanding an upgrade in their computer system. DFW Tracon's next scheduled major computer upgrade was still eighteen months away. Controllers said they couldn't wait. Randy Wayson, who represents the National Air Traffic Controllers Association on the National Transportation Safety Board fumed in the *Dallas Morning News* (vol. 141, October 21, 1989), "We're working 1989 traffic with a computer with the speed of 1960."

Congressman Greg Laughlin (D-Tex.) concurred. "[The DFW incident] sure highlights a very serious problem with the capability of the air traffic control system to handle a growing amount of air traffic." Even the top FAA official at DFW agreed that there was a serious problem.

Four months later, the FAA installed an extra computer processor, solid state memory, and new software. The upgrade greatly increased the capacity of DFW Tracon. Although not acknowledging the problem that had struck during the Texas-OU weekend, FAA Southwest Regional Administrator Don Watson offhandedly noted the need to "better accommodate unusually heavy traffic during holiday periods . . . [and] sporting events."

But even the upgrade did not go smoothly. The new A304 software at first refused to work properly. A team of software engineers had to be brought in from Unisys to modify the coding. Installation of the software was delayed for some twenty-one days.

Milt Scherotter says things operate much more smoothly at DFW now. "There have been four or five days," he says, "where there's no question the computer would have bombed again had we not had all this new equipment."

But other air traffic control centers have not received the attention DFW Tracon did, and subsequently have not received updated equipment, even though they have also suffered the same problem.

The General Accounting Office has found that nearly half of Tracons surveyed around the country had experienced similar losses of computer data. At one busy Tracon, the computers crashed right in the middle of the Friday afternoon rush hour. All computer information displayed on the radar screens, including the identity, altitude, and speed of each airplane, simply disappeared. The outage lasted for sixteen agonizing minutes.

The Tracon manager reported that heavy traffic had simply over-loaded the computer and caused it to shut down. Air traffic con-trollers managed to avoid any accidents by responding the same way DFW controllers had: by holding all departures and keeping all incoming flights away.

At another busy Tracon, it was noted that computer information would randomly vanish from some screens, only to mysteriously reappear several minutes later. The pattern would then be repeated. FAA officials reported this had happened on several different oc-casions, and would last for fifteen to twenty minutes at a time.

Several Tracons reported difficulties characterized by a host of more minor, but still potentially serious, glitches. Radar display screens would flicker. The flashing would occur as the computer was struggling to update its information. The flickering often re-sulted in what GAO called "confusing and obscured information being presented to controllers." During periods of heavy air traffic, the response time of the computer would slow down. The computer would ignore air traffic controllers' attempts to update or delete information about planes. Controllers would have to make several attempts to delete old information. The report noted the danger inherent in having the controllers distracted from their primary air traffic control responsibilities.

The General Accounting Office flatly warns, "computer capacity shortfalls at some large, busy Tracons are impairing controllers' ability to maintain safe separation of aircraft." Why are the FAA's computers so often overloaded? The problem is that the FAA often fails to adequately test its air traffic control software programs to make sure they will not overload computers.

All new software programs are tested at the FAA's Technical Center in Atlantic City, New Jersey before they are distributed to be used at air traffic control facilities. But the Technical Center does not provide a fair test of the software. Although many Tracon fa-cilities have to support twenty display screens for controllers, the Technical Center uses only eight. Critics charge the Technical Cen-ter provides an unrealistic test of how software will perform in the real world.

FAA officials claim the Technical Center can install dummy switches to simulate traffic conditions far heavier than those encountered at the busiest Tracon. Still, the GAO noted, "software that was successfully tested at the Center, when placed in an operational facility, does not always accommodate the traffic loads or operate as efficiently at all locations. For example, when one Tracon attempted to use an updated version of software that had been successfully tested, controllers regularly experienced sector loss." Information would simply vanish from radar screens.

Meanwhile, the strains on the system are growing. In trying to reduce the danger of mid-air collisions, the FAA recently added a new requirement that even small private planes must carry computerized altitude reporting equipment when flying close to the nation's busiest airports. In coming years, GAO estimates this will add forty four thousand new airplanes that the ARTS IIIA computers will have to track. By dealing with one safety problem, the FAA is threatening to create another by pushing its overstrained computers even further.

The GAO report concludes, "current system overloads caused by heavy workloads may be increased by [these] additional requirements . . . This would further impair controllers' ability to maintain safe distances between airplanes."

There are other safety problems the FAA has to constantly contend with in its software. An air traffic control computer does not necessarily have to go haywire for it to cause problems. A flaw can simply mean that whoever programmed the computer assumed the people using it would also be error-free.

On May 29, 1987, a United Airlines flight sat on the ground at O'Hare International Airport in Chicago waiting to depart. Air traffic controllers accidentally assigned to it a radar code that should have gone to an American Airlines flight. The United flight took off without incident. A few minutes later, the American flight took off. Controllers assigned it the correct radar code, but the software would not accept the code because it recognized the duplication. But the software had not been programmed to signal an alert in that case, so instead, it simply rejected the American flight as an imposter and ignored it. Controllers tracking the radar blip left by United Flight 441 ordered it to turn left, thinking it was actually American Flight 637. The American pilot, unaware that controllers did not know his actual position, responded to the order. In doing so, he turned directly into the flight path of an inbound American jet, missing it by only five hundred feet.

The FAA is quick to point out that software errors and computer failures have never caused a crash. Airline flights are still the safest way to travel between two cities. But as congressional investigator Gregory Simon pointed out in *Newsweek* (vol. 115, January 29, 1990), "we're just lucky that the first major software accident happened in the telephone system, rather than in the air traffic control system."

When asked if the FAA's software woes could cause accidents, the GAO's Joel Willemssen told *Computerworld* magazine (vol. 24, March 5, 1990), "Certainly it could. It's nothing to be taken lightly."

Even when such problems pose no safety threat, they can still wreak havoc on travellers' plans, and cost airlines money because of delays and cancellations.

Examples are numerous:

- October 19, 1989. Just five days after the Texas-OU incident, the DFW Tracon suffers more computer problems. One of its four computer processors is shut down for maintenance, but then refuses to restart. Flights in and out of the Dallas-Ft. Worth area are delayed for most of the day.
- August 5, 1988. Air traffic controllers at Boston Center switch on their new IBM computer, only to discover a software error is causing the computer to randomly send critical information on flights to the wrong air traffic controllers. Engineers attempt to switch back to their old computer, only to discover it also is not working properly. Controllers decide to continue using the troubled new system until the old system can be repaired. During the next two and a half hours, fifty flights along the

East Coast are delayed, some by as much as an hour.

- February 13, 1988. A computer that tracks flights, routes, and altitudes, fails at the Jacksonville Center. All flights to and from North Carolina, Alabama, and central Florida are halted while the information is transferred between airport control towers by telephone and teletype. More than eight hundred arrivals and departures are delayed for up to an hour.
- November 26, 1987. Flights are delayed over a five-hour period when a computer fails at Los Angeles Center. The malfunction comes during the morning rush, delaying flights throughout southern California and parts of Arizona and Nevada.
- March 9, 1987. Flights at Atlanta's Hartsfield International Airport, the nation's third-busiest, are delayed over a three-hour period when a computer fails.
- February 24, 1987. The Houston Center computer goes down for four hours. Flights in southern Texas and western Louisiana are delayed for up to ninety minutes.
- August 6, 1986. The main computer at Chicago Center fails in the middle of the afternoon rush. A component that transfers information on the crowded airspace around the world's busiest airport from the radar to the computer fails. The backup computer is activated, but the computer is slower and less efficient, causing hundreds of flights to be delayed.

The effects of these disruptions can be widespread. A problem in England affected travel worldwide for days. An error in a new program at the London Air Traffic Control Center in the summer of 1990 caused all computer information on flights to disappear from radar screens.

The error lasted only for about two hours. But its effect quickly spread far beyond London. When the problem began, the LATCC immediately issued a ban on takeoffs of any London-bound flights on the entire continent of Europe. Incoming planes were "stacked up" over Heathrow Airport. Some had to wait ninety minutes before landing.

Even after the error was fixed, air traffic throughout Europe was hopelessly snarled. The long-haul division of British Airways had its schedule disrupted for several days afterwards, because aircraft were now scattered across the globe, far from their intended destinations. Flights had to be cancelled or rescheduled in order to move aircraft to their proper locations.

Such problems may become more frequent as the skies become more crowded and the FAA's computers strain to cope with rapidly increasing air traffic. The National Research Council in October 1990 estimated the number of passengers flown in the United States by the year 2000 will reach one billion passengers a year, double its current level. By the year 2040, traffic could triple from what it is today. The Congressional Office of Technology Assessment warns the nation's air traffic control system carries the potential for a systemwide safety breakdown. To avoid it, OTA suggests that air travel may have to be limited.

The strains on the system are already showing. After President Reagan fired striking air traffic controllers in 1981, concerns were repeatedly raised about the safety of the air traffic control system. Those concerns reached a fever pitch, when in 1987, fliers took note of a seeming rash of near-midair collisions involving jetliners. In early 1987, the number of reported near-misses rose an alarming twenty-nine percent. Recorded errors by air traffic controllers jumped seventeen percent. In the first seven months of the year alone, 272 near-collisions involving commercial airliners were recorded.

The public outcry reached its peak when President Reagan himself nearly became a victim. A Piper Archer, being flown by an AWOL Marine, came within 150 feet of Reagan's Marine One helicopter as it flew towards the Presidential ranch near Santa Barbara, California. Presidential spokesmen said there was no danger to the President; but the near-miss was widely reported.

The Federal Aviation Administration, which has responsibility for air traffic control, was stretched to the limit. The firing of controllers during the 1981 controllers' strike had sharply depleted its staff and, at a time of rapidly growing traffic, had left the air traffic control system both understaffed and inexperienced.

And while air traffic continued to grow, the FAA was also facing the 1990s hopelessly outgunned, armed with 1960s technology. In a world of high-tech semiconductor computer chips, the FAA was still the nation's largest user of old-fashioned vacuum tubes; their inventory of old IBM and Univac computers was outdated and suffering from an alarming number of breakdowns.

The FAA's answer to both problems lay in buying more computers. Advanced computers, it hoped, would permit it to handle great volumes of air travel with fewer manpower requirements. Automation, FAA officials thought, would be their savior.

In 1981, Congress approved an ambitious FAA undertaking. The National Airspace System (NAS), weighing in at a huge $8 billion, would be the nation's largest non-military technology project since the Apollo moonshot. What the FAA sought was nothing less than a total revamping of the nation's air traffic control system, which would serve the growing needs of the nation's transportation infrastructure well into the twenty-first century.

The $3.6 billion order that kicked off the project in 1988 marked the largest contract the Department of Transportation ever awarded, and the biggest job IBM ever received.

By the mid-1990s, the FAA hopes to replace virtually every air traffic control computer and radar display around the country. The controllers of the future will have the vital information needed in front of them on a single twenty-inch-square color display. With safety concerns paramount, federal officials stress the new system will make controllers better able to cope with potential disasters, be they man-made or the whim of Mother Nature. "The system will enable controllers to more easily spot weather hazards and sort flight data," the FAA boasted. "The new computers will be able to plot flight plans from start to finish, assisting controllers in projecting potential problems and redirecting aircraft around them."

As is nearly always the case when federal agencies computerize, the FAA stressed the enormous cost savings it would gain. The new software would enable individual controllers to handle twenty percent more flights. The reduced need for staffing alone, it figured, would save the government $4.5 billion over the lifetime of the system.

The system also offered a financial bonus to the airlines, and presumably, the travelling public. Computerized flow control software would coordinate departures and arrivals from coast to coast, substantially reducing delays. The FAA put the cost savings of increased airliner efficiency at up to $60 billion during the system's twenty-year lifespan.

But even as the FAA proudly made its announcement, the project was already in trouble. The kickoff contract was already one year late. The FAA had already pushed back the completion date for the project from 1992 to 1995.

Almost immediately after the project was approved, the FAA discovered it had greatly underestimated the complexity of the project by optimistically banking on software that was well beyond what was currently available at the time.

Assoc. Director of GAO Kenneth Mead blasted the FAA's procurement procedures. "Many of these [advanced computer technologies] had to be invented, and the FAA underestimated how long it would take to invent some of these things."

The problem is mostly one of software. The FAA had placed too much faith in the ability of computer programmers to develop the complex programs needed to make the sophisticated computers work. Tom Kingsfield, a staff member on the House Transportation Appropriations Subcommittee, bemoans, "That's been the big problem. In virtually every one of the major programs where there's been a delay, it's been a software delay."

The National Airspace System is enormously complex, placing unprecedented demands on its software. As congressional staffer John Shelby explains, "Think of it. You may have eighty, one hundred planes in the air at one time that you're tracking in each sector. And you have to keep them separate from flocks of geese, and from non-commercial aviation, and maybe you've got a hailstorm out there, plus you have to make sure you don't have false signals, and then make sure everything is completely accurate, virtually to the foot. It's just an enormously complicated problem. I don't think anybody in the whole country realized the extent of the problems that would come in trying to make this thing work.''

The FAA's schedule has now been pushed back up to five years. Of twelve major systems, only one has been completed so far. Given the difficulties, the FAA now concedes the project may not be completed until at least the year 2000.

Alarmed by the rising delays in the NAS project, FAA officials in 1987 scrambled, and put together a project called the Interim Support Plan. A number of stop-gap measures, they hoped, such as buying new solid-state memories, computer processors, and additional radar screens, would buy them some time by allowing them to patch their existing computer system together until NAS was finished.

The Interim Support Plan meant that the FAA had to go back to Congress for an additional $415 million in funding. This temporary fix has fallen three years behind schedule. GAO evaluator Sara Magoulick told *Aviation Week & Space Technology* (vol. 133, October 1, 1990), ''The program is slow and behind schedule. The projects fit, but they are not coming on fast enough to solve the problem.'' In at least one instance, delays in parts of the Interim Support Plan proved fatal.

On December 3, 1990, Detroit Metropolitan Airport was struggling. An irritatingly steady stream of snow and sleet had given way to fog banks that obscured the airport. Controllers cursed. The weather had been up and down all day. Landings were banned; visibility was less than the FAA's requirements. But takeoffs, an action where less visibility is required, were continuing, albeit at a reduced pace.

Northwest Flight 1482, with forty-three people aboard, was about to take its place in line. The DC-9 muscled itself back from its departure gate with a powerful blast of reverse jet thrust. Capt. William Lovelace settled in the left-hand seat in the DC-9 cockpit.

This was only Lovelace's thirteenth flight in the last five years. He had just returned from a medical leave of absence for treatment of kidney stones.

Metro Ground Control cleared 1482 from Concourse *A* to Runway 3 Center for takeoff. Lovelace was instructed to follow Taxiway *F*. With a nudge of jet thrust, he steered the jet away from the terminal using the small tiller to the left of his seat. On his lap, he held a chart of Metro Airport's labyrinth of taxiways and runways. He would be relying on the charts and on signs along the airport grounds to guide him. Visibility outside the plane was murky: Lovelace could see barely four hundred yards in front of him.

Within seconds of leaving the terminal, Lovelace accidentally turned on to the wrong taxiway. Lovelace and co-pilot James Schifferns spotted their error as they watched a taxiway sign go by, and notified ground controllers.

Up in the tower, controllers couldn't see any of this, owing to the low visibility. For all practical purposes, the controllers were blind and were forced to rely on what the pilots were telling them as the airplanes crisscrossed the airport grounds.

Controllers told Lovelace to use an alternate route, Taxiway *X*, to reach his takeoff runway. The pilots, guiding their plane slowly through the fog, came to an intersection where they were to turn onto Taxiway *X*. But, instead, they accidentally turned onto the runway that was in use at the time.

By now, the pilots were unsure of precisely where they were. They could not see that ahead of them in the fog, Northwest Flight 299, a 727 with 153 people aboard, had just been cleared for takeoff, and was on a collision course with them.

Controllers asked Lovelace to verify where he was. "We're not sure," came the reply. "It's so foggy out here. We're completely stuck here."

Ground Control sensed trouble. "Okay, well, are you on a taxiway or a runway or where?"

"It looks like we're on 21 Center [the active runway] here," radioed back the Northwest DC-9.

The controller keyed his mike and issued a warning to Northwest 1482: "If you're on 21 Center, exit that runway immediately, sir."

In the tower, shouts filled the air. "Stop all aircraft! Stop all aircraft!" a supervisor yelled frantically.

A controller turned pale. "I have cleared an aircraft," he reported.

Northwest Flight 299 came barreling down the runway. The 727 had nearly reached takeoff speed, and was accelerating through one hundred miles an hour. As Lovelace and Schifferns watched in horror, the plane's glaring lights appeared out of the fog an instant ahead of the plane itself. The planes struck with a sharp, fatal blow.

The 727's right wingtip pierced the smaller DC-9 just below the right window of its cockpit, slicing open a seam along the plane's fuselage the entire length of the plane, peeling the aluminum skin of the DC-9 back like a giant can opener would have done. At the end of its incision, the wingtip impacted the plane's right engine, ripping it from its mounting, and shearing fuel lines.

The rear of the plane exploded in a shower of shrapnel and flame. Choking smoke and searing fire swept through the cabin "like a wave in the ocean," as one survivor put it. Passengers unbuckled their seatbelts and rushed towards emergency exits.

Aboard the nearly airborne 727, Capt. Robert Ouellette reacted instinctively as his plane's wingtip struck the DC-9. He instantly initiated an aborted takeoff, even as he struggled to retain control of his airplane. He swiftly applied brakes and reverse engine thrust. Spoilers popped up on the wings. Despite the impact and the fact the plane was near the end of its takeoff run, Ouellette, in an incredible display of skill, brought the wounded 727 to a rapid and safe stop. Fuel was gushing from a tear in the plane's wing, presenting a fire danger. Miraculously, no one had been hurt aboard his plane. Ouellette wondered how the people aboard the DC-9 had fared.

The cabin of Flight 1482 had turned into an inferno. The intense fire that began at the rear quickly engulfed two-thirds of the plane. Flight attendant Heidi Joost had been seated in the rearmost jumpseat, astride the two engines. When the fire broke out, she grabbed at the nearby emergency handle. The mechanism was supposed to jettison the plane's tailcone, and provide an exit for herself and passengers seated in the rear of the cabin. The emergency release failed.

Joost and seven passengers died. In the aftermath, investigators blamed the weather, confusing airport ground markings, and the failure of the DC-9 pilots to properly follow directions. But delays in the FAA's Interim Support Plan may also have played a role. Detroit Metropolitan Airport was supposed to have received an advanced computerized ground-tracking radar system that could have prevented the accident.

The system is called Airport Surface Detection Equipment, or ASDE 3. The system is designed to allow controllers to track all ground movements at airports. Even in zero visibility, controllers using the new equipment can see precisely where each plane is on the ground. Coupled with a system called Airport Movement Area Safety System, or AMASS, it issues a computer voice warning whenever an airplane strays onto an active runway.

"It would have provided a warning," says Dave Nussbaum, program manager at Norden Systems, developer of ASDE 3 and AMASS. "In rain or heavy fog, the radar can see through those weather conditions." Nussbaum says it's debatable whether it would have provided enough time to keep the two airplanes from colliding, but notes, "It'll help find an aircraft that is lost [on the ground] due to weather or an aircraft that has strayed into the wrong area."

ASDE 3 was supposed to have been installed at Detroit Metropolitan before the Northwest accident. But the schedule for installations was pushed back twice, falling more than three years behind schedule.

Much of the delay was software-related. Shortly after the contract for ASDE 3 was signed in 1985, the FAA came back and added some new features to the system to make it easier for controllers to use, and to improve the system's capabilities in bad weather. Nussbaum says, "Unfortunately, some of those changes did add development time."

In addition, Norden found it was taking more time to develop the software than it had originally thought. The FAA's original schedule called for 1990–91 installation of Airport Surface Detection System 3 at the nation's seventeen busiest airports. At the time of the Detroit collision, ADSE 3 was in use at only one test site, Pittsburgh International. Norden now says installations will not be completed before 1994.

In the meantime, potential collisions remain a significant safety problem at most major airports. Federal safety statistics recorded an alarming 223 instances in 1989 where an airplane or an airport vehicle strayed onto a runway being used by a departing airplane.

"The potential of a high-speed collision is great," David Haase, spokesman for the Airline Pilots Association, told the Associated Press following the Detroit accident. "We feel there are several areas that need to be addressed to reduce the risk, everything from high-tech surface detection radar to changing the phraseology used by airport personnel."

But the high-tech end of that is still slow in coming.

That development of ASDE 3 has fallen behind schedule is not unexpected. In all, at least eleven of the sixteen projects in the FAA's Interim Support Plan for temporarily shoring up the nation's air traffic control system have been delayed. Items such as additional radar screens, new computer disk drives, backup power systems, long-range radar transmitters, and airport weather sensors, have similarly fallen behind schedule. The Interim Support Plan, designed as a temporary measure to bridge the gap between the current air traffic control system and development of the National Airspace System may not be completed until 1998. That would be six years after the NAS itself was originally supposed to have been completed and on-line.

Meanwhile, the NAS plan keeps getting pushed back. It may not be completed until sometime in the next century. Owing to the numerous delays, the cost of the NAS has climbed steadily. By the time the FAA reported back to Congress in 1987 on the status of the project, six years after its start, it had grown from $8 billion to a massive $16 billion. The General Accounting Office now estimates the project could cost $24 billion by the time it's finished. Congressman Bill Lehman (D-Fla.) asked incredulously, "Are we really going to have to spend $24 billion?"

Lehman probably didn't want to hear the answer he got back from GAO Assoc. Director Kenneth Mead: "At least that much."

Jack Buck is manager of the FAA's Technical Center in Atlantic City, New Jersey and is responsible for testing the FAA's new software. He admits the FAA is hampered by federal procurement regulations that limit the amount of money that can be spent on drawing up requirements and technical specifications for new systems. Time and money spent here greatly reduces the number of software problems encountered down the road, but budget policies generally do not favor the concept of spending a little extra money early in hopes of saving more money later. Buck says, "We don't put enough engineering in upfront before we award the contract. So there's schedule problems sometimes, and the amount of software code that's needed is underestimated."

As stated previously, the FAA had made the mistake of being overly optimistic about the enormous task of developing the amount of software the NAS would need. In developing one computer system for en-route air traffic control centers, IBM found the system required sixty-two percent more lines of code than originally thought.

As Mead explained, "The bulk of the delays can be attributed to a serious underestimation by the FAA of the technical complexity of the different technologies and the interdependencies of the plan. It's like having a three-pound bag and trying to put five pounds of bricks in it."

FAA officials admit now the NAS plan was not thoroughly thought out in advance. "It's a significant technical challenge," FAA Assoc. Administrator Martin Pozesky says, "It has to be done in a very highly reliable manner with error-proof software. The time it takes is not an indication of trouble."

Secretary of Transportation James Burnley blamed "abstruse

and cumbersome'' federal procurement rules with limiting his authority, and saddling him with delays that caused equipment to be obsolete by the time it was installed. By the time IBM replaced its old 9020 computers with new 3083s, the next generation of IBM computers, the 3090s, had been available on the market for over a year.

The FAA defends the project and its backbone, the Advanced Automation System, the key to modernizing its air traffic control computers. The AAS makes up nearly half the projected cost of NAS. FAA Associate Administrator Pozesky says, ''The AAS is a sound economic investment. The program will allow us to go forth in the most cost effective way.''

But critics say the FAA did not spend enough time examining competing systems. For the terminal control portion of the Advanced Automation System, the General Accounting Office charges that a different design could have saved $750 million.

The GAO also says that in selling the program to Congress, the FAA used unsound methods that overstated the NAS plan's benefits. The FAA claims its modernization plan will generate $7.3 billion in savings benefits to airlines and passengers. Reduced delays, says the FAA, will mean savings in fuel consumption and maintenance costs. But the lion's share of the savings, according to the FAA, will be a $4.2 billion benefit from savings of passengers' time. If the flow control software saves five minutes on a three hundred-passenger flight, the FAA figures society gains fifteen hundred minutes, or twenty-five hours. But the FAA calculates each passenger values his/her time at a generous $25 an hour. As GAO Deputy Director Daniel White explains, ''A lot of people question how accurate that is.''

Further, seventy-one percent of the $4.2 billion savings would be realized by passengers ''gaining'' less than fifteen minutes. White says, ''We question the value placed on [such] small passenger time savings.''

Critics also question the FAA's promises that NAS will allow it to handle increased traffic with smaller increases in the number of controllers. As Rep. James Oberstar, chairman of the House Public Works Subcommittee on Aviation, told *Aviation Week & Space Technology* (vol. 133, August 6, 1990), "All these grand, bravado statements . . . that we will have a stronger system, that we can do more with less . . . We know now that you don't do more with less."

Of even greater concern is ensuring the safety of the new software that will guide our planes through the skies. The FAA is still trying to figure out how it will certify that the software will work both safely and reliably under all conditions. Software for the NAS will be a great deal more complex than today's system. The current ARTS IIIA uses a mere sixteen thousand lines of software code. The new AAS system, by comparison, will use over two million. The system will never be completely error-free. As Jack Buck, who heads up the FAA's software testing says, "There will be errors in the software. The main object of testing is to make sure there's not a critical error in the system."

The AAS is scheduled to undergo rigorous testing to try and ensure it will not have any serious errors that could compromise flight safety. The new computers and software will undergo ten months of simulations, where they will be given real data from air traffic control centers. After that, they will be run through another seven months of field testing handling actual air traffic, alongside the current system.

But the General Accounting Office worries that the FAA's plan for developing AAS software "does not adequately mitigate technical risks and does not provide for suitable operational simulation of the advanced automation features."

"Certainly with a program as complex as this," Buck admits, "it's difficult to establish what it takes to certify the system. It's a complex problem." Indeed, any software faced with the myriad of combinations of planes, skies, and weather faces a daunting task. Mike DeWalt, an FAA resource specialist working on a flight management computer system for airplanes, notes, "The sky is home to an infinite number of aircraft velocities and positions." In other words, anything is possible, and the scenarios the software will be required to cope with are endless.

Its many technical problems notwithstanding, the FAA is proceeding with the National Airspace System, convinced it will be the answer to the nation's rapidly expanding appetite for travel. "The NAS plan is a journey, not a destination," former FAA Administrator T. Allan McArtor said. According to one congressman, NAS will lay "the foundation for a whole new decade of dramatic change for the United States. Air travel has gone from something available only to the executive to being available broadly to the public. There's been an extraordinary increase in the number of people using air travel."

McArtor correctly points out that the nation's air traffic control system is operating safely, despite delays in developing NAS and rapidly growing traffic. "The air traffic control system continues to handle unprecedented traffic levels safely. Delays have been decreasing significantly."

In truth, delays have been increasing as the FAA struggles to keep up with increasing traffic while saddled with outmoded equipment. Cautious FAA managers are often slowing down departures as a safety precaution to ensure safe workloads for controllers. The number of delayed flights rose eight percent in the first half of 1990, compared with the year before.

The problem is especially severe at major airports that are already overcrowded, including New York, Chicago, San Francisco, Atlanta, and Dallas-Ft. Worth. Delays of flights in and out of New York-area airports jumped forty-five percent in the first six months of 1990. Airline officials are worried the system is falling further and further behind. Robert Baker, senior vice president of operations for American Airlines, predicts in *Aviation Week & Space Technology* (vol. 133, August 6, 1990), "The FAA will continue to make investments in people, training, and equipment, but it won't be able to keep up with the rate at which aviation is growing through this century." Baker predicts delays will increase, and more airports will develop serious congestion problems.

Controllers continue to wait for the new equipment they hope will allow them to efficiently handle the traffic. But between development delays, cost overruns, and finding a way to ensure utmost safety, the plan to prepare the nation's air traffic control system for the 1990s and beyond continues to struggle. The National Airspace System remains more than four years behind schedule.

A congressional staffer says wearily, "Should the FAA have anticipated these problems? Sure they should have. We wish they had." But he points the finger at the complexity of the FAA's undertaking. "It's the damn technology."

Chapter Eight

THE GOD OF WAR

THE military, because it always insists on the latest state-of-the-art, high-tech gadgetry, is constantly running afoul of software problems. The Defense Department is the largest single user of software. With budgets forever being limited, the Defense Department follows a policy that constantly crosses new frontiers, and frequently pushes the state of the art past its limit. The Pentagon would rather buy a smaller number of highly advanced weapons than buy a much larger quantity of less sophisticated weapons.

Operation Desert Storm—the war against Iraq in early 1991— showed how devastatingly effective such weapons can be. Laser-guided "smart bombs" hit bridges and enemy communication centers with remarkable accuracy. Cruise missiles—small, unmanned bombers—weaved their way through downtown Baghdad at blinding speed, flying underneath Iraqi defenses and hitting their targets with pinpoint precision. The successes racked up in the Gulf War convinced many Pentagon planners that the high-tech weapons approach is a sound one.

But undermining those successes is a long record of cost overruns, extreme delays, and disappointing performances by some new weapons. The problems can often be traced to software difficulties.

A 1989 Pentagon study admitted that trying to get computer software to work right in high-tech weapons systems is a constant headache. Fully ten percent of the Pentagon's budget, some $30 billion annually, is spent on software, which is riddled with errors. "Since operational software is complex," the report noted, "it usually contains design flaws, and these are hard to find and often painful in effect."

Particularly painful for one test pilot. When the Navy's F/A-18 supersonic jet fighter was developed, its advanced fire control computers worked as they were supposed to, sort of. When the test pilot pushed the button to fire his missile, the computer opened the launch rail, releasing the missile. It then ordered the missile to fire, and then closed the launch rail. The only trouble was, the computer closed the launch rail too quickly, before the missile had a chance to completely drop away from the airplane.

The missile's rocket engine ignited with the missile still firmly attached to the airplane's wing. The startled pilot was suddenly a passenger in the world's largest pinwheel. The plane careened wildly through the air, and dropped twenty thousand feet in seconds before the missile ran out of fuel and the pilot was able to regain control.

At Warner-Robins Air Force Base in Georgia, the Air Force owns an advanced radar system it has to turn off every time planes take off and land. The Air Force is afraid the system could cause planes to accidentally fire their missiles or jettison bombs and fuel tanks. The PAVE PAWS radar system is designed to watch for submarine ballistic missile attacks against the eastern United States. The impressive structure stands ten stories high in the shape of a pyramid. PAVE PAWS can reportedly detect an object the size of beach ball from fifteen hundred miles away.

The trouble is, PAVE PAWS does more than simply detect missiles. The electronic impulses it sends out are so strong they could endanger Air Force planes flying nearby. Brig. Gen. Burton Moore acknowledges that a 1987 Air Force study shows PAVE PAWS radar could set off tiny explosive charges called EEDs, Electro-Explosive Devices. The tiny explosive bolts do everything from firing ejection seats and air-to-air missiles to releasing bombs and external fuel tanks. The bolts can be triggered by high levels of electromagnetic energy.

The north face of Warner-Robins AFB's PAVE PAWS must now be turned off every time a military plane comes within one mile of the facility. The Air Force commissioned a study (at a cost of six hundred thousand dollars) to see what could be done about the problem. The Air Force proposed either installing a device that would automatically shut down the radar when planes approached (which would cost $27 million), or else just up and move the entire facility ($37.7 million for that option).

The Air Force claims the radar is only shut off for brief periods. During that time, General Moore says, "There is no loss of warning because of our space-based early warning system." If General Moore is correct, then why does the Air Force need the new radar system in the first place?

The truth is, these are not isolated examples by any means. Even Pentagon officials admit they have a tough time finding software that works properly, and doesn't cost taxpayers enormous sums of money. Even the pride and joy of the U.S. Air Force is a prime example of how costly and troublesome Pentagon software can be, and how it undermines our nation's ability to defend us.

PILOTS will enthusiastically tell you what a joy it is to fly the B-1B manned bomber. On a typical mission profile, the B-1B will streak just above the nap of the earth, skimming at blinding near-supersonic speeds at altitudes so low it seems to threaten the tops of high trees as they blur by. The trees remain unscathed. The advanced terrain-following radar electronically sweeps the ground immediately in front of the speeding plane. The detailed images of the terrain are fed to the computerized autopilot, which makes minute adjustments to the controls, keeping the plane steady at an altitude of only two hundred feet above the ground.

Ultra-sensitive sensors in the nose search out pockets of air turbulence. Before the disturbance even reaches the tail, the autopilot has already made adjustments to the controls, maintaining a smooth ride for the cockpit crew. The variable-geometry wings constantly sweep forward and back, adjusting the wing profile to suit the flying needs of the moment.

At altitudes high overhead, enemy fighters use their "look-down/shoot-down" radar, which searches in vain, trying to lock on to a radar image that is hopelessly lost amidst the clutter of trees and hills. On the ground, radar operators at air defense stations catch only a fleeting glimpse of the bomber as it speeds at low altitude underneath their radar net.

Aboard the B-1B, the electronic warfare officer notes the presence of enemy radar. He selects the appropriate ECM (Electronic Counter-Measures) jamming, and sends out "chaff" and "music"—jamming signals that reduce ground-based radar to a hopeless jumble of confusing signals.

Once near its target, the B-1B releases its payload, unleashing a barrage of nuclear-equipped cruise missiles or short-range attack missiles. The B-1B heads for home, having successfully penetrated a sophisticated array of enemy defenses.

Given its promised capabilities, it is little wonder President Reagan pulled the once-cancelled project off the shelf in 1981, dusting it off and making it a central part of his defense build-up of the 1980s. Most members of Congress did not argue that the country badly needed a new manned bomber to replace the aging fleet of B-52s. What did bother Congress was the potential price tag: upwards of $200 million per plane. In order to persuade Congress to give the go-ahead, Air Force officials promised there would be a $20.5 billion cost cap on the entire program. To ensure there would be none of the infamous cost overruns that had plagued other programs, the Air Force itself would serve as prime contractor on the project, guiding the development of the bomber and shepherding it through production.

Many congressman were skeptical, but reluctantly decided to go along.

True to its word, the Air Force delivered its fifteenth B-1B on schedule in September 1986, within budget, and proudly declared the bomber was now an operational part of the Air Force. The first B-1B squadron was commissioned three months ahead of schedule. A relieved Congress went back to Capitol Hill to tend to other duties, as the Air Force set about deploying its new airplane. It was almost too good to be true, many members of Congress thought.

In 1987, rumors began circulating around the Capitol. Disquieting reports the B-1B didn't work properly. Rumors of pilots constantly shutting off faulty equipment. Congressional investigators rushed back to the B-1B project like firefighters hurrying back to a second alarm at a fire they thought they'd already put out. What they found was an expensive airplane that was riddled with what a congressional report called numerous "immediate" and "particularly acute" problems that kept the B-1B short of its expected performance, and limited the survivability and mission effectiveness of the plane. The problems were mostly software-related.

As a flying machine, the B-1B works fine. Pilots say it flits through the skies like a hummingbird. One B-1B pilot describes it by saying, "Comparing flying the B-1B against the old B-52 is like comparing a Porsche with a pickup truck." The B-1B flies with both uncanny precision and disarming ease.

The problems mostly lay in the software for the plane's sophisticated computer electronics, without which it simply cannot perform its mission effectively. The B-1B is designed to penetrate enemy lines by swooping in at high speeds at extremely low altitudes.

The electronic heart of the B-1B is a supposedly state-of-the-art system called the ALQ-161. Its advanced software does several things at once. The B-1B's airframe is studded with ultra-sensitive receivers that can detect radar signals, even if the signals are too weak to detect the B-1B's presence. The software processes the information at high speed, and automatically determines the most effective way to "jam" or otherwise confuse the radar signal. The ALQ-161 will automatically select one of several modes of ECM and begin broadcasting signals from the B-1B that render enemy radar useless. People on the ground operating defense radars simply see confusing random patterns on their radar screens instead of targets.

But the ALQ-161 did more than just jam enemy radar: it jammed itself. Certain jamming frequencies strongly interfered with the system's own sensitive radar receivers. Following a three-month study, the Pentagon determined it would have to buy a new radar receiver for the plane.

The ALQ-161 also broadcast signals that jammed the plane's radar altimeter, preventing the pilot and the computer autopilot from receiving an accurate reading of the plane's altitude. Likewise, the ALQ-161 interfered with the bomber's target acquisition radar, requiring pilots to shut off one or the other.

Worse, instead of hiding the plane from enemy radar, the ALQ-161 accidentally sent out signals that would help enemy radar find it. In essence, the ALQ-161 sometimes became a "beacon" sending out "Here I am" radar signals, destroying any advantage of surprise.

The development of the system had been rushed. The Eaton Corporation's AIL Division, which had been awarded the contract to build the system, was producing units even though developers hadn't finished designing them. Investigators found there were several different "ALQ-161s," depending on whether a B-1B had rolled off the assembly line early or late.

In the spring of 1988, *fully a year and a half after the first B-1Bs became operational.* Eaton-AIL finally agreed to conduct a series of tests to prove the ALQ-161's worthiness. The airplane's ECM equipment proved embarrassingly easy to trip up. Air Force officials simply employed a standard electronic warfare countertechnique that any enemy would use: while the plane flew around the test range, they bombarded the B-1B with high-powered radar signals to confuse the equipment. The super high-amplitude radar signals drove the ALQ-161's computer processors into overload, keeping its software from operating successfully.

Eaton officials complained in *Aviation Week & Space Technology* (vol. 129, October 10, 1988) that the test was not fair. Charles

Barron, president of Eaton's AIL Division groused, "Those power levels will drive anybody's receiver into the nonlinear area [overload]." Such an excuse would hardly hold up during a war. (One calls to mind *The Far Side* cartoon where inside the circle of covered wagons under attack from Indians, a beleaguered settler looks at the flaming arrows streaming in, and complains, "Hey! Can they do that?")

Not to worry, though. Eaton promptly proposed a series of fixes that would upgrade the system and make it work as promised. The cost: up to $1 billion in additional funding.

Two and a half years later, in October 1990, the Air Force wheeled out the B-1B for a follow-up series of tests. This was now *four years* since the B-1B had become operational. But tests conducted at Eaton-AIL's defensive avionics integration facility in Deer Park, New York showed the ALQ-161 still would not work as promised. Technicians found certain combinations of potential enemy threats that the system would numbly refuse to react to.

The other half of the plane's electronic dynamic duo was to be its Terrain-Following Radar. This computer marvel was essential: it would allow the pilot to zip around at low altitudes by electronically "seeing" the contours of the terrain ahead of him, even at night and in bad weather. Trees, hills, valleys, buildings; all could be avoided, but only with Terrain-Following Radar.

The TFR was not ready by the time the first B-1B rolled out. It was a year behind schedule and even when it was finally ready, could not fly in certain types of bad weather or on all the mission profiles promised. The Air Force had simply not realized how difficult it would be to develop the software programming for such a complicated system. Instead of hugging the ground and following the contours of the terrain, the TFR, according to one government report, "erroneously flew up then down and did not accurately follow the terrain."

There were other problems as well. The B-1B carried internal diagnostic computers that would tell mechanics where problems lay, greatly simplifying maintenance. But the CITS (Central Integrated

Test System) gave three times more false alarms than correct alerts. On some flights, the CITS was generating two hundred false alarms of equipment failure. The problem: faulty software programming.

The Tail-Warning System was designed to protect the crew from missile attack from behind the aircraft. But the TWS suffered from false alarms, or else did not warn crews of incoming missiles.

On top of all this, the B-1B came in overweight. The Air Force tacked on a hefty eighty-two thousand pounds to its combat weight, a heft no diet could cure. The resulting design change limited the B-1B's effectiveness by cutting its ability to fly low and fast and by reducing the amount of fuel and armament the plane could carry. The Air Force's solution was to add another computer. The delay: minimum two years.

The B-1B's computers were overloaded. The software programs had taken up more of the memory space in the computers than had been anticipated. A new computer system with added memory capacity would have to be added. Cost: another $414 million.

Members of Congress were outraged. Charles Bowsher, the U.S. Comptroller General said simply, "We took a big risk on the B-1, and we lost."

Congress and the Department of Defense thought they had forged a new and unique partnership. In exchange for a "hands-off" policy from Congress that would eliminate what Pentagon planners derided as "bureaucratic micro-management" of the project, the Air Force promised to produce an effective weapon with no cost overruns. The Pentagon promised a fully capable weapon. In July 1982, the Under Secretary of Defense for Research and Engineering promised the plane would have survivability, be able to attack at low levels, maneuver close to the ground, have full electronic counter-measures capabilities, and fly in night and all weather. "That will be the airplane that will be *delivered from day one* [author's emphasis]."

In point of fact, the Air Force's B-1B project office knew as early as 1982, only months after the project started, that the project was experiencing serious problems. The main culprit was a lack of time for proper testing of its numerous software-driven components. In order to meet the $20.5 billion cost cap they had promised, B-1B managers knew they had to produce bombers at the most economical production rate, four per month. To reach that goal and still produce the first squadron by 1987, the Air Force set an impossible development and production schedule. The schedule had no room for delays or any unexpected problems. Key computer components, such as the ALQ-161, TFR, and CITS were put into production bombers, even though their software did not work properly, and their subcontractors had not had time to complete development and testing phases.

In the end, many different "B-1B's" were produced, depending on where on the schedule they came off the assembly line. Eventually, the Air Force announced it would have to spend an additional $387 million to backtrack and standardize each plane. Without that, the Pentagon would have been forced to stockpile an enormous amount of space spare parts for all the different versions of the plane.

Even as B-1Bs rolled off Rockwell International's assembly lines in Downey, California with faulty systems, Air Force personnel continually assured members of Congress that the problems were being fixed. In 1985, the director of the Defense Department's testing office said the plane was experiencing a few difficulties, but "attainment of the full operational capability is ultimately achievable."

Secretary of Defense Casper Weinberger took great pains to explain what a close watch he was keeping on the program. "Every two weeks, we review the B-1B program. Are you on time? Are you within budget?" The trouble was, officials were fixated with controlling the cost of the project, and nobody stopped to ask if the systems aboard the B-1B actually worked.

Ironically, the problems only came to light when some Pentagon officials began asking for additional money to buy even more B-1Bs. Advocates of other Air Force programs, competing for limited budget dollars, began pointing out reports of inadequacies in the B-1Bs.

The official budget line shows the B-1B program came in totalling $20.4 billion, within the $20.5 billion cap the Pentagon had promised. But the Air Force had shifted items totalling $2.6 billion, such as simulators, spare parts, and maintenance equipment, to other parts of the budget to hide them.

The Air Force eventually stepped forward and said it needed money for additional testing and "enhancements." In budget jargon, enhancement means you're buying additional capabilities. In the case of the B-1B, critics argue the enhancements are nothing but fixes for flaws in the original plane.

The Office of the Secretary of Defense itself was not even kept abreast of the problems in the B-1B's development. The director of Pentagon Operational Test and Evaluation, in charge of monitoring weapons systems, was questioned about that in March 1987. When asked whether the Secretary of Defense was getting accurate information on the B-1B's lack of progress, the response was not reassuring.

Rep. Les Aspin: In fact the Secretary was not getting candid information.

John Krings: That is a fact . . . No, I don't think they [the Air Force] were completely candid, no.

Krings testified he "knew about the problems with the B-1 from the beginning," but never told Congress until after the problems became common knowledge. Rep. Charles Bennett (D-Fla.) complains, "Congress never received information which would have alerted us to the serious problems with this program."

Almost every B-1B pilot you speak with stands by his airplane. They feel the B-1B's software deficiencies have been overblown by the media and Congress. One B-1B pilot says, "If we go to war tomorrow, this is the plane we'd want to go to war in." The pilots have learned to live with the plane's deficiencies. They don't fly as fast or as low as they might, and they shut off systems at various times to avoid interference with other computer systems.

But even if the men who fly the B-1Bs have found ways to fly around the shortcomings, the fact remains that, as is the case with so many other highly complicated weapons systems, the Pentagon simply didn't get what it paid for.

In April 1988, the one hundredth and final B-1B bomber was delivered to the Strategic Air Command. Less than six months later, a GAO investigation concluded that despite the Air Force's success in delivering the airplanes on time, "the performance of the aircraft was considerably less than originally intended . . . [which] necessitated operational restrictions and prevented some training."

Ninety-seven B-1Bs are now operational in the Strategic Air Command. Planes are still being retrofitted with new equipment to bring them up to their original design specifications, and systems are still being tested for flaws. Over $10 billion has been spent so far over the original cost estimate of $20.5 billion. Correcting the deficiencies on the B-1B, along with some improvements, could cost an additional $9 billion. As Charles Bowsher, U.S. Comptroller General says, "We really didn't test that plane in some of the key components of it until it was built. And so we have a plane today that has some limited capabilities, and we had tremendous cost overruns on it."

The B-1B's limited capabilities were borne out by its conspicuous absence from Operation Desert Storm. The air war involved virtually every aircraft in the United States arsenal, from F-117A Stealth Fighters to A-10 Thunderbolt Tank-killers. But while these planes were streaking across the skies over Iraq and Kuwait, all ninety-seven Air Force B-1Bs were sitting on the tarmac at Strategic Air Command bases in the United States. The Air Force claims the bombers had to stay at home to provide nuclear deterrence against the Soviet Union.

But many members of Congress were outraged. When they were first approached to fund the B-1B program in the 1980s, they were assured by Pentagon officials the plane would be fully capable of fighting both a nuclear and conventional war. In October 1988, Air Force Maj. Gen. Michael D. Hall assured members of Congress that "the B-1B is fully capable of performing the penetration mission that was promised."

But in the Gulf War, the Air Force instead used its old, reliable fleet of B-52s, the last one of which was manufactured in 1962. Rep. John Conyers, Jr. (D-Mi.) says the absence of the United States's most modern operational bomber was because the plane was simply incapable of carrying out its mission.

The Air Force claims seventy percent of its B-1B fleet is mission-ready at any given time. Conyers disputes that, saying the actual fully mission-capable rate is zero. "The Air Force has admitted to me in writing," he says, "that not one aircraft is fully capable of doing the job it was designed for. This is primarily due to software and embedded computer failures."

Despite the successes achieved by many of the Pentagon's new weapons in the Persian Gulf, Conyers warns the B-1B is still a dangerous example of what can go wrong in developing new weapons systems. "In an age of incredibly complex, astronomically priced arms procurements, the B-1B must become a lesson for the Pentagon on how not to buy a weapons system. The same weak procurement safeguards and controls which caused the B-1B fiasco are present in virtually every large weapons system purchase we make."

Another weapons system to suffer from this type of rushed development is the Navy's new advanced attack submarine, the SSN-21 Sea Wolf. It is being built with an untested fire control system to handle its weapons. Its AN/BSY-2 combat computers will contain one of the largest software programs ever written for a submarine. The computers will hold a whopping three million lines of code. The Navy is rushing ahead with development of the SSN-21 Sea Wolf even though the computer system could have a difficult time meeting the Navy's requirements. Asst. U.S. Comptroller General Ralph Carlone warns, the Navy is taking "significant risks with the shortcuts that are being taken in the software testing area." The General Accounting Office claims the Navy could end up with software that is "unstable." The fire control system, which both tracks enemy targets and controls the submarine's weapons, could operate unpredictably and produce different results, even when operated under the same conditions.

But the Navy is in a hurry to bring the Sea Wolf on-line. The submarines, which run $1.1 billion apiece, are already behind schedule and could run $16 billion overbudget. In addition, many of its components do not work as promised. In the current program, fifteen submarines will already be coming down the assembly line, at a cost of over $20 billion, before the Sea Wolf even undergoes its first sea trials. Incidentally, the Navy is asking for funding to add two more of these new submarines to the program.

The C-17 transport plane, DDG-51 Burke-class destroyer, Mark-50 torpedo, TF-45 jet trainer, AN/BSY-1 Army combat system, and AMRAAM air-to-air missile are similarly scheduled to be operational before testing is completed. Many of these projects have already reported problems in their software.

Software is becoming a huge part of the cost of new weapons systems. The software alone for the Army's proposed Cherokee Light Helicopter Experimental program for a new generation of scout helicopters will cost an estimated $35 billion. In Europe's new European Fighter Aircraft project, software takes up fully half of the plane's development costs.

But despite its growing importance, software, it often turns out, is the neglected stepchild in developing weapons systems. As the DoD's Defense Science Board admits, "we cannot get enough of it, soon enough, reliable enough, and cheap enough to meet the demands of weapons systems." The DSB points out software reliability has now become "a critical problem."

An official in the Office of the Deputy Director for Defense Research and Engineering told the GAO that seven out of every ten new weapons systems are having major problems with software development, and the rate is increasing.

Democratic Sen. Ernest Hollings of South Carolina notes, "I have heard time and again about software problems that have delayed the development of weapons systems. As the systems get more complex, the bugs in the software have more places to hide and are more expensive to find and fix. Sometimes it takes millions of dollars and several years to get things working the way they are supposed to. The B-1B avionics system is one of the most glaring examples."

SEN. Al Gore (D-Tenn.) is one of the leading experts on computer technology on Capitol Hill. "In recent years," he claims, "there have been hundreds of weapons systems crippled by bad software." The B-1B, it turns out, is a classic example, but far from the only one. In November 1990, a report by the House Government Operations Committee concluded, "The Air Force's B-2, the Army's Apache AH-64 helicopter, and the Navy's SSN-688 Los Angeles-class attack submarine have all experienced major problems integrating computer technology into the weapon system . . . [causing] major program cost overruns and schedule slippages." The report also concluded the weapons simply did not work as promised once they were delivered.

Why is software becoming the Achilles' heel for weapons systems? Software, the numeric codes that instruct the computer in what to do, is susceptible to what computer people call "requirements creep." Pentagon managers generally do a poor job of defining what they want the software to do. Even after a contract is awarded, project managers are forever going back to the planners and saying, "Oh by the way, we need it do this, and we also need it to do that."

Within the complicated military bureaucracy, far too many people can add to a wish-list of what a new weapon should do, but few people can take things off. The result is overly complicated weapons that are often beyond the present state of the art. The Defense Science Board refers to this "too-many-cooks-spoil-the-broth" syndrome as "the presence of too many intermediaries between the ultimate user and software specifier," which makes weapons, "too complex for the mind of man to foresee all the ramifications."

U.S. Comptroller General Charles Bowsher explains, "They tend to start off as being designed to be a Chevrolet and they end up being designed more to be a Cadillac, and then we find out the Cadillac doesn't work too well."

The problems of military software are more managerial than technical. Project managers are often career officers with little technical or computer background, who end up befuddled trying to grasp the complexities of software. A Navy captain can find himself commanding a destroyer one year, and then suddenly managing the next-generation missile development office, with only a short break in between to attend the Navy's program managers development school. Heading up a project development office is a reward for displaying impressive managerial skills, not for any sort of technical expertise.

As Col. Charles Fuller, director of Air Force mission-critical computer engineering, explains in *Aviation Week & Space Technology* (vol. 132, January 15, 1990), "They're doing the job, but they're making a lot of mistakes doing it because of a basic lack of understanding of policies, regulations, software engineering, [and computer] languages and hardware."

The result, Comptroller General Bowsher says, is often short-changed software testing. "I think lots of times, when the [software] tests don't go too well, people say, 'Well, let's keep going. We can straighten it out later.' That's where the [systems review] people ... should be very rigorous."

Developing software for the military means having to navigate a bureaucratic maze. There are a bewildering assortment of more than 250 different standards and policies in the Defense Department for developing or buying software. There is no one Pentagon office that is in charge of overall software policy. Instead, dozens of different offices all claim different responsibilities for software, depending on the project involved.

The results are all too often predictable. Fuller complains, "I have yet to be associated with a software program that came in on cost and on schedule. It's a long-term problem that's going to take time to fix. And until we get it fixed, we're going to keep facing the same kinds of problems we're seeing now."

Gen. Bernard Randolph, former commander of the U.S. Air Force Systems Command, noted wryly to *Software Engineering Notes* (vol. 15, January 1990), "We have a perfect record on software schedules—we've never made one yet, and we are always making excuses."

Sen. Al Gore of Tennessee claims, "Computer programs containing a million lines of code are becoming commonplace in the military now. Good hardware, of course, is next to useless without good software."

As military software grows more complex, it becomes impossible to predict when computerized weapons will be tripped up. In 1982, during the Falkland Islands War, a British frigate was sunk because the commander was making a phone call. An Argentine Air Force jet had fired a French-made Exocet missile at the ship. At the same time, the captain of HMS *Sheffield* was placing a radiotelephone call to naval headquarters in London. As bad luck would have it, the radiotelephone was using the same frequency as the homing radar for the Exocet missile. The phone call prevented the ship's electronic counter-measures from detecting the incoming missile. Twenty crewmen were killed.

At a demonstration of the proposed Sergeant York mobile anti-aircraft gun, computer sensors aboard the Sergeant York reportedly detected sunlight flashing off the brass decorations of Army generals, and turned its turret towards the suspected "targets." Members of the reviewing stand dove for cover. Luckily, mechanics had installed special movement locks that prevented the guns from being fully aimed in that direction.

Software is now proliferating in every aspect of the military. The use of computers in the military is growing exponentially. While in 1985, the Pentagon spent $9 billion on what it terms "mission-critical" software weapons systems, that number ballooned to $30 billion by 1990. Even relatively simple weapons such as tanks that once carried no computers are now stuffed with laser-guided aiming systems and navigation computers.

During the Vietnam-era, jet fighters such as the popular F-4 Phantom carried virtually no computer software. The F-16D, designed in the mid-1970s, holds about 236,000 lines of computer code in its various systems.

By contrast, the new B-2 Stealth bomber holds over two hundred computers, using an enormous 3.5 million lines of software code. And the needs for computer software aren't just in the air. In order to keep flying, the B-2 requires ground support equipment for maintenance and testing that uses an additional eighteen million lines of software code. Systems like the B-2 simply cannot function without an enormous mass of software.

Advanced planes such as the B-2 bomber, F-117A Stealth fighter, and the Air Force's proposed ATF fighter plane literally could not even fly without computers. The planes have become essentially flying computers. Their shapes, designed to deflect radar waves, are aerodynamically unsuited for flying. The machines cannot stay in the air without computers making thousands of tiny corrections to the controls every second. If the computers fail, as has reportedly happened with some of the Stealth fighters, the plane becomes unstable and literally flies itself apart within a matter of seconds.

Pentagon planners envision that even individual infantry soldiers will soon carry a number of computers. Hand-held navigation computers will draw signals from satellites to calculate a soldier's position to within fifty feet, and then relay that information to computers at field headquarters, allowing commanders to electronically track troop movements. Enclosed helmets will detect the presence of toxic gases and automatically darken visors when laser beams strike. Soldiers will have computer chips embedded in their teeth that will carry emergency medical information and personnel data. These electronic ''dog-tags'' will be virtually indestructible.

But proper development of software for all these gadgets often seems to take a back-seat. The military is littered with incompatible software. A government report noted that in 1974, the Defense Department was using over three hundred different and largely incompatible computer programming languages, ''making it difficult to move programs among computer systems and expensive to maintain these programs.'' As an example, the Navy and Air Force's A-7E jet attack fighter is a relatively simple piece of equipment. It contains a bare thirty-seven thousand lines of software code for its various computers. But to reconfigure the plane to carry a new missile costs $8 million, because of the need to match up the software in the missile to the software in the plane.

The military *is* making an attempt to standardize software in its weapons systems, as well as make it cheaper and more reliable. The Pentagon in 1975 contracted with CII-Honeywell Bull of France to develop a new simplified software language. The result was Ada, named after Countess Augusta Ada Lovelace, daughter of Lord Byron. The countess was also the developer of the first computer language. Her new namesake program was designed from the start to be quicker to write, easier to debug, and cheaper. In an effort to try and standardize all of its software, the Pentagon in 1983 declared that Ada would be used for all "mission-critical" software. A report prepared by Carnegie-Mellon University heralded the arrival of Ada as "a major step forward in addressing Department of Defense software development problems."

Ada has shown some promise. Raytheon, a leading defense contractor, claims Ada has cut its cost of developing new defense software by up to fifty percent. Other contractors report similar results.

But in practice, Ada has been a far cry from the much hoped-for solution to the Pentagon's software woes. As the Congressional Office of Technology Assessment notes, the Pentagon hasn't even been successful at developing the equipment it needs to use the new language. "Early compilers of the new Ada language have been so slow, unwieldy to use, and bug-ridden that they have been worthless for real software development."

Ada software programs are, generally, large and unwieldy. Programs run slowly in cases where speed is critical. In the Cameo Bluejay Project, for instance, the Army wants to use fast, nimble scout helicopters to scour the frontlines of a battlefield. Using high-speed computers, the helicopters will search, detect, track, and electronically jam enemy weapons.

But in order to achieve these objectives and avoid attack, the computers must detect the presence of enemy weapons and then respond within four milliseconds, or 4/1000ths of a second. A General Accounting Office report notes, though, that "Ada has not worked well in critical applications that require precise timing."

The March 1989 report concluded that Ada's difficulties could be overcome, but that it would take several years of further development.

The Pentagon has begun to recognize the problem of software development. Secretary of Defense Dick Cheney in 1990 ordered senior officials to draw up a software master plan for overhauling the DoD's software policies. The report recommended that a new high-level office be created that would have responsibility for setting policy for all Pentagon computer software.

As Dr. George Millburn, deputy director of Defense Research and Engineering said (*Aviation Week & Space Technology*, vol. 132, April 16, 1990), "Software is such a big problem that we have to solve it. We have no choice."

But at a conference of defense industry software experts called by the Pentagon to review the plan, critics warned that if the new office merely becomes another layer of bureaucracy, it will not solve the Defense Department's software problems.

THE problems the Pentagon faces in dealing with software take a quantum leap when considering the massive Star Wars defense plan, also known as the Strategic Defense Initiative. The vision of a ground- and space-based defense that would knock enemy warheads out of the sky before they reached the United States was championed by President Reagan and the cause has been picked up by President Bush.

The notion is a grand one: an array of ground-based lasers, orbiting rockets, and particle beams would blast incoming ICBMs before they could threaten the United States.

In his State of the Union address of 1991, Bush seized on the success of the Patriot anti-missile missile in the Persian Gulf to push for stepped-up research for SDI. The Patriot, hastily dispatched to

Israel and Saudi Arabia to protect lives and equipment, was the high-tech hero of the Gulf War. It claimed a success rate of better than ninety percent in intercepting incoming Iraqi Scud missiles. Defense Secretary Cheney remarked, ''all you have to do is watch the Scuds over Tel Aviv and Riyadh to have a sense of the extent to which ballistic missile capability is a threat to U.S. forces.''

But the Patriot's system is far less sophisticated than SDI's would have to be. The Patriot is basically a short-range defense system with a maximum range of forty miles. The incoming ICBMs the SDI system would face would need to be intercepted at a much greater distance, hundreds of miles away, before they re-entered the earth's atmosphere. Also, while a Scud-A flies at a speed of Mach 4, or four times the speed of sound, an incoming nuclear warhead enters the atmosphere at more than fifteen times the speed of sound.

SDI would easily be the most complicated construct ever developed by man. But it faces a number of daunting challenges. The system must:

— detect and track thousands of incoming targets and separate them from thousands of decoys and other space satellites (in contrast, the Patriot system can detect only up to one hundred incoming missiles)
— direct a series of particle beams, lasers, and anti-satellite missiles in a precise manner to destroy all targets
— be protected from attack itself by nuclear and laser weapons
— work perfectly the first time it's used (against an incoming barrage of tens of thousands of nuclear warheads, a mere ninety-five percent success rate—as achieved by the Patriot— means that hundreds of warheads would successfully strike their targets)

Anything less than virtual perfection would result in, as the Congressional Office of Technology Assessment puts it, ''an unacceptable number of warheads reaching their targets.'' A polite way of saying that many U.S. cities would still end up being vaporized.

SDI will require controlling software up to one billion lines of code long—by far the most complicated software ever written. Such software is far beyond the current state of the art. Even to build programs that will simulate SDI on the ground will require software technology that does not exist.

In addition, the software would require constant upgrading as potential enemies develop new technologies to try and thwart SDI. Such constant changes would threaten to introduce even more errors into the programs. A government advisory panel made up of leading industry experts concluded such a system could never be properly tested. "No adequate models exist," the panel concluded, "for the development and testing of full-scale ballistic missile defense systems. There would always be irresolvable questions about how dependable the software was."

Testing out the system in a computer-simulated war game simply wouldn't do the job. In short, the only way to find out if the system works correctly would be to hold a nuclear war. Many leading computer software experts, including Prof. Frederick Brooks of the University of North Carolina feel that successful SDI software is within the capabilities of current software technology. But the Office of Technology Assessment was pessimistic when it reported back to Congress that even given "extraordinarily fast rates of research, there would be a significant probability that the first, and presumably only time the ballistic missile defense system were used in a real war, it would suffer a catastrophic failure."

And even though proper software is critical to the SDI effort, it has been largely ignored. The Department of Defense's own Defense Science Board advised, "The Strategic Defense Initiative has a monumental software problem . . . no program to address the software problem is evident."

COMPUTER software problems are not limited to weapons programs. The Department of Defense spends $9 billion a year on automated information processing systems to handle everything from payroll to the monitoring of project costs. But John Caron who, as Asst.

Commissioner of the General Services Administration, provides technical advice to all federal agencies, including the Defense Department, notes that DoD information software systems have a horrible batting average. "Less than one percent of large systems," he says, "are finished on schedule, within budget, and having met all user requirements."

In 1988, the House National Security Subcommittee looked into reports of cost overruns and huge delays in the Navy's new automated financial computers. The proposed Standard Automated Financial System was needed to correct accounting deficiencies that were known to exist at fourteen Navy Research, Development, Test and Evaluation Centers. The department also wanted to standardize its accounting procedures for its Navy Industrial Fund.

By the time government investigators looked into the matter, STAFS was tremendously overbudget. Worse, the project was running five years behind schedule, and Navy accountants said they no longer wanted to use it.

The subcommittee dug deeper and found that not only was the STAFS project significantly overbudget and late, but seven other DoD automated information projects were having similar trouble. All told, the eight systems were $2 billion over budget. Six of the projects were behind delivery schedule, with delays of between three and seven years. As a congressional report noted, "very few instances can be found of large [DoD] automated information systems projects delivering the capabilities and benefits promised on time and on budget."

One problem is that Pentagon planners do a horrendous job of estimating costs. A House subcommittee report cited an "almost total lack of accuracy in cost estimates." Congress was told that one particular financial accounting system would cost $6 million. When the contract was signed, the Navy said the price was $32 million. By the time budget planners pulled the plug on the project, $230 million had already been spent, and the total estimated cost had risen to $500 million.

When costs grow out of hand, the military has a curious way of dealing with the problem. Instead of raising the bridge, the Pentagon lowers the water. Told that a spare parts distribution computer system was about to go $23 million over budget, the Pentagon responded by ordering only thirteen of the twenty-three it had planned to purchase. The system was supposed to save the Air Force $3.1 billion by improving inventory control. Because of the system's reduced size, savings totalled only $193 million.

Troubles with automated information software are nothing new. Back in the mid-1970s the Defense Department found it had major problems in both the development and acquisition of automated data processing. Millions of dollars were wasted, and planned systems fell years behind schedule. Many times, there was little or nothing to show for the time and money invested. The House Government Operations Committee concluded in November 1989, "Continual cost overruns and excessive cost growth of these systems . . . appear common and chronic."

Ironically, the Defense Department had instituted a cost control program as early as October 1978. MAISRC, the Major Automated Information System Review Council, was to control costs and manage development of automated information systems. Analysis shows MAISRC did a good job of keeping programs on cost and on schedule. MAISRC reviewed programs regularly to make sure they were proceeding as planned.

But a gap existed because MAISRC often did not review software, concentrating instead on development of the equipment itself, rather than its programming codes. Software for these projects was often never adequately tested. Forty percent of information system contracts reviewed by the Defense Department had not been subjected to DoD software quality control checks. One Navy program had been fed "dummy" data for its test run, rather than actual numbers that would have reflected the type of information the computer would have to deal with once operational. And instead of making the computer tackle a myriad of different calculations, testers simply had the computer perform the same calculation ten thousand times. The testers said the computers would have crashed if a more realistic mix of data reflecting real-world problems had been used.

Often, project managers simply find a way to skirt regulations. A General Accounting Office investigation found that, on one computer modernization project, half of the project's software did not have proper controls that would have avoided common development problems. On an accounting project called ETADS, project managers told GAO they did not have to do a feasibility study because it had been waived. The officials were unable to find any documents to back up their claim and could not recall who had approved the waiver.

Air Force officials said they viewed feasibility studies, project management plans, and performance reviews as drudgery—work they only did because it was necessary in order to get funding from Congress.

But ironically, although the Pentagon spends so much time and attention on its new weapons systems and the cost overruns of those systems, its biggest failure may be its inability to develop computer-based accounting systems that would, among other things, help catch the overruns before they run rampant. The lack of such systems has left the military with an appalling lack of financial controls over its budget. In other words, nobody's minding the mint. The failure has left the Pentagon unable to tell where all of its money is going.

In February 1990, the General Accounting Office released an audit of the Air Force. The GAO found the Air Force could not account for much of its budget. Fully seventy percent of all Air Force assets were considered "unauditable." The report said the Air Force did not know the true cost of most of its $275 billion in weapons systems, and that most of its financial data was inaccurate and unreliable. Democratic Sen. Joe Lieberman of Connecticut growled, "I found it startling. They apparently don't have in place the kinds of accounting systems that most small businesses consider to be routine." Records were even found to be altered or "adjusted" when figures did not add up properly. "The Air Force has no systems," the report stated, "to accurately account for billions of dollars invested in aircraft, missiles, and engines. When the financial systems produce information that is obviously wrong or merits investigation, these problems are often ignored.''

The Air Force has an annual budget of over $90 billion. It dwarfs the largest businesses, and even most other agencies in the federal government. It has over nine hundred thousand employees, and outstanding contracts of more than $250 billion. As U.S. Comptroller General Charles Bowsher said, "We're not talking about small money here, and we're not accounting for it properly."

In March 1990, the GAO found the Pentagon had stockpiled more than $34 billion in unneeded items. That's four times greater than the entire inventory for General Motors. One dismayed senator groused, "Looking into the Pentagon supply closet is like opening the door to Imelda Marcos's closet." The Navy had stockpiled a supply of a particular machine tool to last it thirteen thousand years at the present rate of use. The Navy had also squirreled away fifty-three reactor assemblies for a sonar system, even though it only needed two. The report rightly called it a waste of tax dollars.

Unable to keep track of what it owns, it's not surprising that the Air Force found that supplies such as munitions, firearms, sleeping bags, and even F-16 jet engines were being stolen and sold or "fenced" for personal profit.

The Air Force, the GAO found, simply lacks good accounting and records software. "Air Force financial systems," according to the GAO, "are unquestionably incapable of providing the kinds of reliable financial information every organization needs for effective and efficient management." Even though the service maintains a bewildering assortment of 131 different accounting and financial management systems, it lacks even a basic double-ledger accounting system, found in most businesses, that could help it keep track of the costs of its weapons.

Carryover from the military buildup of the 1980s is as much to blame as a lack of good accounting computer software. As Sen. James Sasser, a Democrat from Tennessee, points out, "One of the problems is, they've just been awash in money. They don't have the proper accounting procedures. Maybe we'd be doing them a favor if we cut them substantially and sort of focused their mind on how they're actually spending their money.''

Charles Bowsher, head of the GAO, thinks the Air Force never intentionally misled Congress on the cost of its weapons and inventory; it simply didn't have the mechanisms to properly account for its costs. Still, Bowsher notes, ''There are a couple of problems here. I think they finally realize they're going to have scandal after scandal, and reports that are going to point out things that the average American just doesn't think should take place.''

Following the investigation, the Air Force sent a letter to Congress meekly acknowledging that it was aware that ''the Air Force accounting system . . . needs improvement.''

Incidentally, the Air Force was selected for the audit because it was the only branch of the military that had even attempted to keep financial statements in accordance with generally accepted accounting principles for federal agencies. The other branches are in even worse shape.

Chapter Nine

A CASE OF MISTAKEN IDENTITY

As we've seen, the greatest danger from software is that it will always contain errors that can never be completely eliminated. But even when it works perfectly, software can still cause serious problems. Software design may fail to factor in the possibility of human error. The computer can then greatly magnify the inevitable errors people are going to make while using the computer. As Nancy Leveson, who teaches at the University of California-Irvine's Information and Computer Science Department points out in *Software Engineering Notes* (vol. 12, July 1987), "We know that humans can make mistakes. People seem much too willing to blame the human instead of the technology that failed to consider the human in its design. I am starting to be leery when I read that the technology did not fail, the human operator did."

Programmers often neglect proper consideration of the weak link in the software: the person who will be using it. As software grows more complex, the possibility for human errors or misuse of the system grows. Often, even the best software turns bad when it falls into the hands of well-intentioned people. Conditions in the real world, it turns out, are rarely, if ever, what the programmer imagined.

The best examples can again be found in government software, some of which is called on to make life-or-death decisions. Faulty software design results in outcomes ranging anywhere from irritating to lethal. People are killed or falsely jailed because of it. The best example may be found in what has been billed as the most complex battle system ever devised.

AEGIS, in classical mythology, was the body armor worn by the god Zeus. It was so impenetrable that, according to legend, enemies were paralyzed in terror at the mere sight of it.

It was only fitting then, that the U.S. Navy chose Aegis as the name of its state-of-the-art defensive weapons system. The Navy hoped its combination of advanced computerized electronics and powerful missiles would protect its ships with a shield that would have made Zeus proud.

Since the mid-1970s, the Navy was searching for a way to protect its carrier groups from an onslaught of new enemy missiles that could attack the fleet from great distances. Crews had to worry about threats that could sink them without ever even being seen.

Capt. Will Rogers had faith in the Aegis. As commander of the cruiser USS *Vincennes*, he had every confidence it would protect his ship and the 357 men under his command. Rogers had reason to be confident. Aegis was, at the very least, an impressive system, taking up a huge portion of the 566-foot-long ship, and making up nearly half of its incredible $1.2 billion cost. Navy officials proudly touted the system as "the shield of the fleet," guardian of its other ships.

Aegis's powerful radar could detect any plane flying within 200 miles of the ship. The system could detect and track hundreds of different targets, separating commercial air traffic, military planes, surface ships, and submarines. Aegis's radar is reportedly sensitive enough to detect a basketball 150 miles away. Its ESM, Electronic Support Measures receivers, can discriminate between hostile aircraft and harmless commercial air traffic. In case of attack, the Aegis computers would direct the ship's defense. The ship carried an impressive array of missiles, antisubmarine rockets, torpedoes, and guns. The Aegis software would automatically assign a weapon to each target, depending on the threat it posed. If attacked from the air, Aegis could fire a deadly volley of up to eighty missiles into the sky in only four minutes.

But Aegis had a troublesome history. Adm. Thomas Davies had tried to have the program killed, a nearly impossible task in the military. Unsuccessful, Davies retired in frustration, calling Aegis "the greatest expenditure to get the least result in history." Davies was right about the cost. Aegis cost $600 million for each of the ten ships it had been installed on.

Navy brass had stood by the system. Aegis was, according to then-U.S. Navy Secretary John Lehman, "the most carefully tested combat system ever built." Before the first Aegis system had been commissioned, it had been subjected to over one hundred thousand hours of testing, and three trials at sea. Lehman constantly brushed aside suggestions from the media and Congress that tests had revealed flaws in Aegis.

In truth, Aegis had failed miserably in its first two sea trials aboard the USS *Ticonderoga* in 1983. No one came away impressed except the Navy and RCA, the prime contractor of the system. Rep. Denny Smith (R-Oreg.) contends the results of the tests were classified for the simple reason that the results were not favorable to the Navy. When Smith finally succeeded in seeing a copy of the test results, he discovered the page of the report with the test scores was mysteriously missing. Congressional sources report Aegis had hit only six out of twenty-one targets. Unlike Zeus's impenetrable armor, the Navy's Aegis could hardly be counted on to protect ships at sea.

Angry members of Congress demanded a new round of tests. This time, Aegis did perform almost flawlessly. The Navy was more than happy to declassify these results. Of eleven test drones, or unmanned aircraft, launched against the target ship, Aegis shot down ten. What the Navy failed to reveal was that the operators of the system had been told the path and speed of the drones before the "surprise" attack took place.

Even more importantly, the rigorous testing failed to reveal one tiny, fatal flaw in Aegis's software, one a House subcommittee would later term "a seemingly trivial design decision."

None of that mattered to Capt. Will Rogers of the *Vincennes*: he had faith in the Aegis system and the men under his command who were operating it. The thirty-three-year Navy veteran was at home at sea, and was clearly in awe of the impressive technology surrounding him. He once described the sleek *Vincennes* to a friend as "the most sophisticated ship in the world, bar none" (*Newsweek*, vol. 112, July 18, 1988). Concerns about the Aegis system were not on the mind of Captain Rogers. But what did bother Rogers was the ship's mission.

"HELL, Toto, this ain't Kansas," read the sign on Capt. Will Rogers III's desk. It sure as hell wasn't Kansas. Rogers would just as soon have been somewhere else. The date was July 3, 1988, and the *Vincennes* was cruising the waters in the eastern end of the Persian Gulf. Since May, the ship had been part of a Navy flotilla of twenty-nine vessels in the Middle East. The mission was simple: keep the Persian Gulf and the critical Strait of Hormuz open to sea traffic.

The task was not nearly as simple as it sounded. The seven-and-a-half-year-old war between Iran and Iraq was still raging, and the hostilities now threatened American lives and American national security interests. A year earlier, after a U.S.-managed oil tanker in the gulf was struck by an Iranian missile, President Reagan had ordered the Navy into the gulf. Kuwait had threatened to cut off its shipments of oil to the rest of the world rather than risk its ships to further violence. Kuwaiti supertankers were temporarily reflagged as American vessels, and the U.S. Navy was assigned to provide them escort through the treacherous gulf.

It was a set-up many Democrats in Congress were not happy with. Neither were Navy officials thrilled about sending thousands of servicemen into a raging war zone, all for the purpose of protecting America's flow of oil.

But orders were orders, and it was the job of Captain Rogers to carry them out. Few men were better equipped for it than he was. Raised deep in the heart of Texas, the landlocked Rogers had somehow grown up with an intense desire that would lead to his commanding the pride of the fleet. He had risen quickly through the fleet by time and again displaying intelligence, leadership, dedication, and bravery. He had worked his way up the chain of command from a decrepit mine sweeper destined for the scrap heap to the destroyer USS *Cushing*. A turn as a fellow at the Navy's prestigious Strategic Studies Group, a spot normally reserved for the elite of the officers' corps, polished him for what was to follow. With only six hundred ships needing command, and seventy-two thousand officers yearning for them, only the Navy's very best earned the sought-after rank of captain. By virtue of having been tapped for command of one of only ten brand-new Ticonderoga-class, Aegis-equipped cruisers in the fleet, Rogers had been selected, as one officer told *Newsweek*, "as the best of the best of the best."

But Rogers would need all his training and intellect for this task. Navy tactics generally favor battle in the open sea. Defending against attacks in the narrow twenty-five-mile-wide Strait of Hormuz was like battling in a bottle. It was too easy for dangerous lightning-quick air strikes and missile attacks to arise from out of nowhere. Just a year earlier, the cruiser USS *Stark* had been accidentally struck by a pair of Exocet missiles fired from an Iraqi fighter plane. Fifty-eight dead and wounded was the result. There was clearly reason to be cautious.

Navy escort ships spent much of their time within easy striking distances of Iranian missile batteries. The *Vincennes* had been assigned to the western mouth of the strait, on the eastern edge of the Persian Gulf. Reconnaissance missions from the aircraft carriers in the group brought back daily progress reports on construction of an Iranian base at Kuhestak, to the north. Once completed, the Iranians would be able to launch Silkworm missiles that could strike anywhere in the strait. Just to the north of the strait, at Bandar-Abbas, Iranians had stationed a half-dozen F-4 fighters, and had recently transferred two F-14 fighters there. Ironically, the F-14 was also the U.S. Navy's top-of-the-line fighter. Navy commanders felt that only the sophisticated electronics and firepower of Aegis cruisers could provide cover for tankers and for other Navy vessels. To the *Vincennes*, then, fell the risky mission of guarding the western sector of the strait.

The Persian Gulf was a hotbed of activity. Two months earlier, the Ayatollah Khomeini, spiritual leader of the Islamic nation of Iran, dared to take on what he called "The Great Satan," namely the United States. Whether planned or irrational, the move was ill-advised. Iran chose to challenge the U.S. and was resoundingly defeated.

Following the collision of the frigate USS *Samuel B. Roberts* with an Iranian mine in the Persian Gulf, President Reagan in April ordered what he called a "light" retaliatory strike against Iran. Navy ships destroyed two oil platforms at Sassan and Sirri in the southern gulf. The oil rigs had been frequently used by Iranian spotters to target oil tankers as they transited the gulf waters. The Iranians responded by attacking U.S. ships with missiles and speedboats. The U.S. Navy, using a sophisticated and overpowering array of ship-to-ship missiles, jet fighters, and laser-guided "smart" bombs, sank or crippled six Iranian Navy vessels, turning back the attacks with relative ease, losing only two servicemen when a Sea Cobra helicopter failed to return. The Iranians had fired seven missiles at various U.S. Navy ships and planes. None hit its intended target. Operation Praying Mantis was deemed a complete success.

State Department and Navy officials realized it would only be a matter of time before the Ayatollah would lash out again. Anti-terrorism officials braced for a possible wave of bombings, hijackings, or kidnappings as a form of retaliation. The only question, it seemed, was not if an attack would take place, but when and where. Asst. Defense Secretary Richard Armitage stated in an interview in May 1988 that effective counterterrorism efforts by the United States had made it more difficult for terrorists to exercise their deadly craft far from their home nations. For that reason, Armitage warned, the next strike was likely to be in the Middle East, close to home.

This July 3 morning, as every morning, Rogers was at his post in the Combat Information Center, deep in the below-decks bowels of the ship. The CIC is a fifty-foot-by-thirty-foot room that forms the nerve center of the massive cruiser. Gone are the days when a ship's captain can stand on the bridge of the ship and see everything that's happening with a pair of binoculars strapped around his neck like John Wayne in *Operation Pacific*.

The CIC is where a captain needs to be to make the correct decisions. All information used to operate and defend the ship flows through this room. The business-like chatter among the men was heavier than normal today. The *Vincennes* had only been on patrol in the gulf for the last five weeks. Just two months earlier, the *Vincennes*, while on her way to the Middle East, had gone through a series of training exercises in Middle East tactics at Subic Bay in the Philippines. The crew practiced how to differentiate commercial air traffic from hostile military aircraft. The ship also fired its surface-to-air missiles twice during the four days of training. But none of the men aboard the *Vincennes*, Captain Rogers included, had ever been in combat.

Topside, it was another typical steamy gulf morning, hot and humid. Scattered clouds dotted an otherwise clear blue sky. Inside, as always in the windowless CIC, there were few clues as to time of day, or what if anything was going on outside.

The room was manned by fourteen crewmen and Rogers. It was their job to manage the Aegis defense system: to monitor for hostile ships and planes, and if need be, destroy them. The room was dominated by four huge computer screens that each measured nearly four feet across. Lighting in the CIC was deliberately kept low to help the crew focus on the readouts. The computer screens provided most of the illumination, casting an eerie bluish glow.

The men were on edge this morning, their senses honed to a keen sharpness. U.S. intelligence had warned the fleet to expect trouble. Clandestine interceptions of Iranian communications suggested members of the Iranian Revolutionary Guard, the same terrorists who had kidnapped fifty-two Americans from the U.S. embassy in

Tehran in 1979, were planning some sort of strike against the U.S. Further, the intelligence reports said, the attacks would be timed to disrupt the Americans' July 4th weekend. And the target would likely be the target situated closest to Iran: the U.S. Navy.

The last three days had been troublesome ones. Iraq launched new air strikes against Iranian oil terminals. Iran had responded by attacking tankers. Captain Rogers had received intelligence reports warning him to "anticipate Iranian ship attacks in retaliation for Iraqi Air Force attacks on Iranian tankers." Just yesterday, the USS *Montgomery* had responded to a distress call from a Danish ship. The *Karama Maersk* was being attacked by three Iranian gunboats firing machine guns and rockets. The *Montgomery* fired a warning shot, the speedboats retreated, and the *Karama Maersk* continued unharmed.

The reports also noted, with special alarm, the redeployment of the F-14 jet fighters to nearby Bandar-Abbas, stating that

> "All units are cautioned to be on the alert for more
> aggressive behavior. The F-14 deployment represents an
> increased threat to allied aircraft operating in the Strait
> of Hormuz, southern Persian Gulf, and Gulf of Oman."

In recent weeks, the Iranians had greatly stepped up their F-14 patrols in the southern gulf. Seven times in the last two weeks, Iranian F-14s were issued radio warnings as they approached too close to U.S. Navy vessels. It had been only three months since two Iranian jet fighters fired two Maverick missiles at a Liberian tanker, just ten miles from where the *Vincennes* was stationed this morning. In the last month, the Navy had issued radio warnings to 150 planes that strayed too close to Navy vessels. One hundred forty-eight of those encounters had been with Iranian Air Force planes; only two of the flights had been commercial.

The crew of the *Vincennes* was expecting trouble this morning, and it found trouble.

To the north, the frigate USS *Montgomery* is on routine patrol in the northern sector of the Strait of Hormuz. A distress call comes across the radio. At about nine-thirty, the crew of the *Montgomery* spots a pack of seven Iranian speedboats heading towards a merchant ship that displays Pakistani markings. Looking through his binoculars, the commanding officer can see the speedboats are sporting manned machine gun mounts and hand-held rocket launchers.

The *Montgomery* radios the sighting to task force commanders. The captain looks again. The force has suddenly grown to thirteen gunboats, split up into three groups. Four of the speedboats take up a position off the port side of the frigate. Over the marine radio, the *Montgomery*'s crew can hear the Revolutionary Guard challenging the merchant ships in the area.

From the deck of the *Montgomery*, crew members can hear explosions to the north. The *Montgomery*'s captain radios task force commanders.

The message is relayed to the *Vincennes*.

"Get Ocean Lord 25 up there," Rogers orders. The *Vincennes*'s lone observation helicopter is on routine morning patrol to the south. The Lamps Mark III is quickly ordered to head north and investigate the gunboat activity.

"*Golf Whiskey, Ocean Lord 25.*"

"This is Golf Whiskey," radios back the *Vincennes*'s Force Anti-Air Warfare Coordinator. "Go ahead."

"*Roger. We have visual contact with a group of speedboats. They are firing small arms at us.*"

"Say again."

"*We can see several small flashes and puffs of smoke about one hundred yards from our position.*"

Rogers immediately gives the order to sound general quarters. A noisy electronic gong interrupts the routine of the ship, and the 357 men of the ship scramble out of their bunks, recreation rooms, and mess halls to man their battle stations. The engine room is given the order for all-ahead full. As the *Vincennes* steams to the north at high speed, crewman uncover the powerful five-inch-diameter gun at the bow of the ship. They load a round of ammunition into the breech, readying the gun for firing.

Within minutes, lookouts spot two groups of small boats closing on the *Vincennes*. The *Vincennes* now takes over tactical control of the *Montgomery*. Rogers orders the *Montgomery* to hold position off the *Vincennes*'s port side. Rogers wants to keep his more powerful ship between the gunboats and the *Montgomery*.

The *Vincennes* fires a warning shot, which is ignored. Rogers radios his superiors. "Request permission to engage speedboats."

"*Confirm speedboats are not departing area*," radios back the USS *Coronado*.

"That is affirmative. Speedboats continue to close position."

"*You have permission to engage.*"

The *Vincennes* opens fire with its five-inch gun. Over the next twenty minutes, the *Vincennes* will fire seventy-two rounds. For every man on the ship, including Captain Rogers, this is their first experience in combat.

As the five-inch guns aboard the *Vincennes* and *Montgomery* begin booming shells at the speedboats, three of the boats break away and begin closing on the larger Navy ships. Lookouts on both ships shout on the intercom that they can see small arms fire coming from the speedboats. Down in Repair Locker 2, deep in the lower decks, crewmen can hear small-caliber shells bouncing off the hull of the ship.

Two of the speedboats are sunk. The crew continues to fire at a remaining speedboat, unaware of the tragic events about to unfold. The surface battle, though deadly in its own right, is a traditional naval battle, using tactics that date back to the days of John Paul Jones.

The second battle that is about to be fought represents modern warfare; it will be carried out by computers and the latest in high-tech weaponry. Neither of the combatants will ever see each other. And it will prove to be far deadlier than the first battle.

While the shoot-out continues on the sea, Iran Air Flight 655 sits on the ramp at Bandar-Abbas airfield on the north end of the strait. Seated in the left seat of the cockpit, Capt. Mohsen Rezayan is annoyed. The flight is twenty-seven minutes late; an interminable wait for a flight this short. From Bandar-Abbas to Dubai is just a short hop to the other side of the Persian Gulf. Once airborne, the flight will take barely twenty minutes. For Rezayan, the trip is what pilots call a milk run—he has made this trip hundreds of times. The seventeen-year Iran Air veteran could practically make this run in his sleep. Despite the official animosity that exists between his country and the United States, Rezayan bears no hatred towards the United States—he had received his pilot's training there, and had lived in Texas for two years. His eldest daughter, Aryanaz, born in Oklahoma City, was an American citizen. As Rezayan's sister-in-law would say later, he "loved and respected the United States and the American culture."

Finally, the Airbus A300 is ready to go. At the request of air traffic controllers, Captain Rezayan sets his radar transponder. The Airbus now begins emitting radar codes known as Mode 3, which are for civilian planes. Mode 2 is reserved specifically for military planes, and the equipment aboard the Airbus is not designed to broadcast on those frequencies. The code 6760 Rezayan enters into the transponder means the Airbus will be identified on radar screens as a civilian plane, and not a military plane. The tower confirms that Flight 655 is emitting the proper signals.

The Airbus lifts off smoothly from Runway 21 and heads out over the Strait of Hormuz. The crew cannot see the battle on the sea below them and fifty miles ahead of them that continues to rage between the *Vincennes* and the Iranian speedboats. The activity is far too small for a jetliner crew to possibly see from the air.

Still, Captain Rezayan knows instinctively that caution is in order today, as it has been for the past several months. The news has been carrying word of new fighting between the United States and Iran in the gulf. Captain Rezayan begins climbing to his assigned altitude of fourteen thousand feet. Rezayan intends to stick precisely to the narrow air corridor that should provide him, and the 289 other people aboard, safe passage over the hostile gulf to Dubai.

Seven minutes to shoot-down. Down in the Combat Information Center, a white inverted *U* appears on one of the four large computer screens, showing an unidentified aircraft has just taken off from

Bandar-Abbas. As the Iran Air flight heads out over the Strait of Hormuz and becomes "feet wet" (overwater), a crewman labels it "UNKNOWN-ASSUMED ENEMY," standard Navy procedure until an identity can be firmly established.

The powerful AN/SPY-1A phased-array radar dish, which dominates the front of the ship, constantly beams out powerful signals that can both spot and identify nearby aircraft. The system is called IFF, Identification Friend or Foe. It electronically "interrogates" aircraft, and waits for their transponder to answer back. By reading its codes, it determines whether the aircraft are friendly and non-threatening, or potentially hostile.

The read-out on the AN/SPY-1A radar shows Mode 3, code 6760, indicating a civilian jetliner. The Identification Supervisor, or IDS, "hooks" the airplane. The radar begins sweeping the area around Bandar-Abbas looking for electronic signals emanating from the aircraft that will help identify it. While the radar is sweeping it picks up a second signal, this time, Mode 2—military, code 1100. The IDS checks his listing of Iranian codes. Naval intelligence had recently broken Iranian codes. The code checks out to an Iranian F-14, known to be based at Bandar-Abbas.

The IDS calls out the information to Rogers.

"Sir, I have a Mode 2, 1100, breaks as possible F-14."

The announcement sends sharp chills down the spines of everyone in the CIC. Several thoughts race through Rogers's mind. The speed-boat attack could be merely a diversion for an Iranian air attack. An F-14 equipped with air-to-ship missiles could easily sink a ship like the *Vincennes*. Rogers knows that if a pilot is to attack the ship, he will probably come in low and fast, descending at the last minute to either fire missiles or even use his own plane as a guided bomb in a suicide mission against the ship. If it's a commercial flight on the other hand, the plane will climb steadily until it reaches its cruising altitude. Whatever it is, Rogers knows he will have less than ten minutes before it will intercept the *Vincennes*.

"Run it again!" Rogers orders.

He wants to be absolutely positive of what he is dealing with.

He knows Bandar-Abbas is home to both military airplanes and commercial jetliner traffic.

"Aye, aye, sir," comes the reply.

The IDS hooks the plane again, and again comes back with the same reading. The reading is confusing: the unidentified aircraft reads back two sets of transponder codes, one civilian, one military. Where the spurious Mode 2 military transmission came from remains a mystery to this day. Weather conditions in the gulf that day were favorable for a condition called "radar ducting," that can carry radar signals far beyond their normal range. Tests conducted afterwards by both the Joint Electronic Warfare Center in Texas and the Naval Surface Weapons Center in Virginia show the *Vincennes*'s radar was ducting that day. Officials speculate the *Vincennes* may have been picking up signals coming from an Iranian military plane sitting on the ground at Bandar-Abbas.

Enough is enough, Rogers decides. He's going to need help in sorting out what in blazes is going on. His ship may now be under air attack. The *Vincennes* sends out an alert to all other ships in the area that it is tracking an unidentified aircraft.

At the northern sector of the gulf, an Air Force AWACS plane is flying a lazy racetrack-shaped holding pattern. A military version of a 707 passenger plane, the AWACS bristles with sophisticated electronics. The AWACS is equipped to sort out the busy gulf air traffic, and can identify and track thousands of separate targets. But the AWACS is too far north, its radar coverage does not extend to the *Vincennes*.

Closer in, the aircraft carrier USS *Forrestal* is patrolling in the Gulf of Oman, just to the east of the *Vincennes*. Upon hearing the *Vincennes*'s call, the *Forrestal* quickly scrambles an E-2C Hawkeye and a pair of F-14 fighters. A smaller, propeller-driven airplane, the Hawkeye is nonetheless crammed with powerful electronics that despite its smaller size, make it very nearly the equal of an AWACS. Had it arrived on the scene in time, there is little doubt the Hawkeye crew would have instantly noticed that the plane taking off from Bandar-Abbas was giving out signals consistent with those of a commercial jetliner. But the Hawkeye, and its F-14 escorts, will arrive too late to save Iran Air Flight 655.

Six minutes to shoot-down. The plane continues on a course leading directly to the *Vincennes*. Although neither Captain Rogers nor his crew realizes it, the ship now sits almost directly underneath the center of Amber 59, the air corridor from Bandar-Abbas to Dubai. Flight 655's path, if uninterrupted, will take it almost directly over the *Vincennes*.

Located amidships, the *Vincennes*'s ultra-sensitive receiver, the SLQ-32, known as Slick 32, scans for electronic emissions from the Airbus. If the weather radar of the Airbus located in the nose of the plane had been activated, the SLQ-32 would have detected it and identified it as coming from a civilian airliner. But in the clear skies over the gulf this day, Captain Rezayan decides there is no need to turn his weather radar on.

The *Vincennes* picks up a second aircraft, this time an Iranian Navy P-3 Orion to the west.

The *Vincennes*'s radio operator challenges the P-3.

"Attention Iranian P-3 . . . this is the U.S. Navy . . . state your intentions," he radios.

The P-3 pilot responds he is on a search mission and will stay away from the *Vincennes*.

The American-made P-3 was one of many gifts that had been unintentionally left for the Revolutionary Guard by the late Shah of Iran when he was overthrown in 1979. A maritime surveillance plane, P-3s are capable of detecting both surface ships and underwater submarines with great efficiency, and can be equipped with powerful anti-ship rockets and torpedoes.

This particular P-3 could be on a search mission. It could also have hostile intentions. Or it could be providing radar guidance to a different plane about to attack the *Vincennes*. Rogers turns his attention back to the unknown aircraft coming in from the north, from Bandar-Abbas.

Five minutes to shoot-down. Rogers orders the ship's radio operators to begin issuing warnings to the unknown aircraft. The first message is sent out on a military-only distress frequency. Before the next five minutes elapse, the *Vincennes* will issue a total of seven radio warnings.

Incredibly, surrounded by $600 million worth of computers and electronics, Rogers has to order a crewman to start frantically thumbing through an ordinary paperback airline flight guide. Rogers is hoping it will show a commercial flight scheduled for that time. There are over one thousand scheduled flights over the gulf each week. The guide lists Iran Air Flight 655, but the unknown target took off twenty-seven minutes after the listed departure time. The Identification Supervisor concludes there is no match. Meanwhile, the ship is still under attack from the remaining Iranian speedboat.

Four minutes to shoot-down. The aircraft is now forty nautical miles from the *Vincennes* and closing. The *Vincennes* begins broadcasting warnings on the international civilian distress frequency, 121.5 mhz. After the attack on the USS *Stark*, the Navy had issued what is known as a ''Notice to Airmen'' to all Persian Gulf countries, warning commercial flights to monitor the frequency. But in reality, few ever did. Warnings issued by other ships to jetliners that strayed too close to U.S. warships in recent months had usually been blithely ignored.

On the outside deck of the ship, crews are still frantically battling the remaining Iranian gunboat. While the nimble speedboat dances around the ship, a round of ammunition jams in the *Vincennes*'s forward gun. The bridge orders a full-rudder turn at high speed in order to bring a five-inch gun at the rear of the ship to bear on the gunboat.

Inside the CIC, the turn makes the atmosphere even more chaotic. As the ship heels over at a steep thirty-two-degree angle, equipment falls, lights flicker, and crewmen struggle to keep from sliding away from their consoles. The crewmen, even though they are situated deep within the ship, can clearly hear the booming of the guns outside, and small arms ammunition bouncing off of the *Vincennes*'s steel hull. Outside, their shipmates continue to engage in what could be a life-and-death struggle for everyone on the ship.

The CIC's two intercom channels are jammed as several crewmen excitedly try to pass information at the same time. Amidst the confusion, several of the crewmen in the CIC hear chatter on their intercom headsets—something about an Iranian F-14 approaching the ship . . .

Three minutes to shoot-down. The target is at twenty-eight nautical miles. Rogers gets on the radio telephone with his immediate superior, Rear Adm. Anthony Lees, commander of Joint Task Force Middle East aboard the nearby USS *Coronado*. Rogers does not need to get authorization to fire. Under the Navy's rules of engagement in the Middle East, Rogers can fire at will if he feels his ship is under threat of attack. Just the same, he wants to cover himself. Captain Rogers informs Lees the *Vincennes* is tracking an Iranian F-14. Rogers says he intends to fire if the plane continues to ignore warnings and comes within twenty miles of the ship. Lees concurs.

The International Air Distress officer continues to broadcast warnings.

"Unidentified aircraft . . . you are approaching U.S. Navy warship bearing 205, thirty miles from you. Your identity is not known. Your intentions are not clear. You are standing in danger . . . alter you course immediately." No response.

Given the official green light to fire if necessary, Rogers orders the unknown aircraft labeled "INCOMING HOSTILE." The Aegis computers mark the Airbus as a potential target. Up on deck, two SM-2 Standard-2 missiles are automatically pulled on to their launching rails, ready for firing.

Aboard the nearby frigate USS *Sides*, crewmen monitor the chatter between the *Vincennes* and the *Coronado*, and the warnings to the unknown aircraft. Something doesn't add up here, one crewman thinks. As he looks at his scope, he sees the plane is broadcasting commercial signals. The airplane is climbing on a standard commercial heading, which would not be consistent with the profile of an attacking airplane. An attacking airplane would be descending as it approached the ship. He turns to his Tactical Action Officer.

"Sir, I think it's a Haj," meaning a commercial flight across the gulf. The officer ignores him. He would later say he never heard the message.

Inside the *Vincennes*'s CIC, the Combat Information Officer looks again at his display. The target is clearly climbing in altitude. The crewman jumps to his feet. He shouts at Rogers, "It's a possible Comair," Navy shorthand for jetliner traffic.

Rogers raises his hand to acknowledge the report, but says nothing.

"Sir, request permission to illuminate target."

The Force Anti-Air Warfare Coordinator wants to power up the ship's twin illuminating radars, which control the missile aiming computers.

"Permission granted." Powerful targeting radars plot an intercept course to the bogey. If the plane is an F-14, the pilot will now hear a musical tone in his headset from his threat receiver, and a red light will illuminate on the front cockpit panel, showing the *Vincennes* has now locked missiles on him. Rogers wants the pilot to know the ship is tracking him. Just yesterday, an Iranian F-14 had taken off from Bander-Abbas and come within an uncomfortably close seven miles of the Navy cruiser USS *Halsey*. The fighter turned away only when the *Halsey* illuminated the plane with its fire control radar. But the Airbus, like most civilian jetliners, carries no threat receiver to warn it of hostile radar.

Up on deck, the aiming computers for the Standard missiles follow their orders and continuously adjust the missiles' aim. As the airplane drifts ever closer, the aiming computers correctly sense the target is increasing in altitude. But the aiming computers do not make tactical decisions, and have a separate set of illuminating radar from that which is being displayed down in the Combat Information Center. Under automatic control, the Aegis will now fire when the plane comes within twenty miles.

Two minutes to shoot-down. The target passes through twenty nautical miles and is still closing. Rogers sits with his hand on the key that fires missiles. He has ordered the system put on manual so he can control it. One simple twist to the "FIRE ENABLE" position will shoot the missiles. Agonizingly, he decides to hold fire. He can't take the chance of shooting down an unarmed aircraft by mistake. As the captain of the ill-fated USS *Stark* had put it, "You're damned if you do, and you're damned if you don't." The lives of every crewman, and the lives of the people in the air, whomever they may be, rest on Rogers's decision. By withholding fire, Rogers is fighting every instinct that told him to defend his ship and his crew. Every scrap of information he has received points to a threat against the *Vincennes*. The aircraft is bearing directly on the *Vincennes*, is reportedly descending, and is ignoring radio warnings to turn away or be fired upon.

He is hoping the aircraft will turn away. The only way of knowing with one hundred percent certainty that the aircraft rapidly closing on his ship was an enemy plane with hostile intentions would be when the plane powered up its fire control radar. As it neared its target, an attacking plane would "paint" the ship with its onboard radar, locking on to the ship as a target for its missiles. The Tactical Information Coordinator reports no such signals are coming from the airplane. But Iranian F-14s were known to fly "cold nose," with their radar switched off. Not only would this avoid detection, but the Iranians were notoriously short of spare parts for their American-made F-14s. The *Vincennes*'s sophisticated electronic warfare equipment, Rogers knew, would pick up the telltale signals when the fire control radar came on, but by that time, Rogers also knew, it would be too late. A targeting radar signal from the plane would mean the pilot was about to fire his missiles.

Communications officers continue to broadcast warnings that go unheeded by the bogey. Although the ship's AN/SPY-1A radar clearly records that the aircraft is climbing, both the Tactical Information Coordinator and the Identification Supervisor say again their computers indicate the aircraft is descending towards the *Vincennes*.

In one more minute, the *Vincennes* will be within the thirteen-mile range of Maverick air-to-ship missiles.

One minute to shoot-down. The target is now at fifteen nautical miles.

Aboard the nearby frigate USS *Sides* there is, according to one crewman, "excitement and yelling about Comair [commercial traffic]." The crewman looks at the Weapons Control Officer's radar display. It shows the plane is commercial traffic, and furthermore, is climbing. Comdr. David Carlson heaves a sigh of relief. The plane is clearly no threat to the *Sides*. He looks at the display. The plane is at eleven thousand feet and still climbing on a standard commercial heading. He turns his attention to the Iranian P-3 that is still aloft, fifty miles to the west.

The Airbus, meanwhile, is still closing directly on the *Vincennes*. Communications gives out one last warning call. The Tactical Information Coordinator reports to Rogers the target is descending in altitude as it approaches eleven nautical miles. Time for a decision is running out. The *Vincennes*'s Standard missiles have a minimum range of six miles. If Rogers waits much longer, the ship will be left nearly defenseless, except for its onboard guns. Outside, the fight with the Iranian gunboat continues. Ten nautical miles, still descending. Rogers thinks back to the USS *Stark*, and the thirty-seven sailors who were killed by two Iraqi missiles when the crew failed to activate that ship's defense system. Many of the sailors died in their sleep while the ship still had the ability to defend itself. Nine nautical miles, still closing, still descending.

Thirty seconds. At 10:53:30 A.M., Rogers gives the command to fire.

Within seconds of Rogers's order, two SM-2 surface-to-air missiles rocket off the deck of the *Vincennes*. The missiles leave their launcher rails just as the target passes through a range of nine nautical miles from the ship. Both missiles strike the aircraft at an altitude of 13,500 feet, eight nautical miles from the *Vincennes*.

Rogers notifies Lees aboard the USS *Coronado* they have just "splashed" an Iranian F-14. He relates that radar indicated the plane was an F-14, was descending on a direct course headed for the *Vincennes*, and had ignored repeated warnings to turn away. Minutes later, Admiral Lees passes the information on to the Pentagon, and also notes that he has just been informed that Iran Air Flight 655 is reported as being overdue in Dubai.

The Pentagon issues a statement that the *Vincennes* has downed an Iranian F-14 fighter. The Navy dismisses Iranian reports that an unarmed civilian jetliner has been shot down. Minutes later, the Navy issues a correction: they have indeed shot down, by mistake, Iran Air 655. Two hundred ninety people were aboard. There were no survivors.

That afternoon, Adm. William Crowe, chairman of the Joint Chiefs of Staff, holds a news conference admitting the Navy's mistake. But Crowe claims the Airbus was outside its assigned air corridor, and was descending towards the *Vincennes* at 520 miles an hour. Only weeks later would the Navy admit the aircraft was in its assigned lane, and was climbing, not descending, at a speed of 400 miles an hour.

Many questions were raised in the wake of the shoot-down of Iran Air 655, chief among them, how could this have happened? How could such a mistake have been made with such a sophisticated computer system at work?

In the weeks that followed, the world came to be familiar with the television news pictures showing horrifying groups of corpses floating in the Persian Gulf. Iran and the United States exchanged blame for the accident. Iranian generals blamed the United States for placing "a missile system under an international airway that is not closed." Admiral Crowe, in turn, blasted Iran for conducting combat operations and then sending a jetliner into a war zone.

As commander of the *Vincennes*, Will Rogers shouldered the blame, resigning his command to retire in San Diego. Rogers's career at sea was over. He resigned by saying, "This is a burden I will carry for the rest of my life. But under the circumstances and considering all the information available to me at the moment, I took this action to defend my ship and my crew."

Computer tapes from the *Vincennes* were immediately whisked off and hand-delivered to the Navy Surface Weapons Center in Dahlgren, Virginia for analysis. Upon playback, Navy investigator

Rear Adm. William Fogarty soon discovered there was a stunning difference between what the *Vincennes* crew reported, and what the computer tapes revealed. The tapes clearly showed the airplane was cruising at an altitude of 12,500 feet, normal for a commercial jetliner, had been climbing, and was not emitting military signals.

The crew had reported the plane was only at 9,000 feet, was descending towards the *Vincennes*, and was giving off both commercial and military signals. In his final report on the incident, Fogarty concluded that Aegis had provided accurate information. The crew had somehow misinterpreted the data.

In truth, the sophisticated Aegis software had fed the crew a bewildering flurry of information that was readily misinterpreted. Both the Tactical Information Coordinator and the Identification Supervisor, the two crewmen responsible for tracking the flight path of the plane, consistently reported to Captain Rogers the plane had leveled off at an altitude of 9,000 feet, and then began descending towards the *Vincennes*.

The operators had fallen victim to the one major flaw of the Aegis, the aforementioned "seemingly trivial design decision."

The radar image of the Airbus on one of the giant computer screens displayed the airplane's position and heading. But critical information about the plane's altitude was omitted, and instead displayed on different consoles. Correlating the two pieces of information proved difficult at best.

Using the computer tapes taken off the ship, top Defense Department officials sat through an eerie replay of the incident in the Navy's Aegis mockup at Wallops Island, Virginia. After watching the fateful events repeated again and again, Joint Chiefs of Staff chairman Adm. William Crowe stated, "I recommend that some additional human engineering be done on the display systems of the Aegis. It seemed to our inexperienced eyes that the Commanding Officer should have some way of separating crucial information from other data. The vital data should be displayed in some fashion so that the CO and his main assistants do not have to shift their attention back and forth between displays."

Defense Secretary Frank Carlucci agreed, "We think it's a good idea to display altitude and range on a large screen. I think you could probably even put an arrow on [showing] whether it's ascending or descending.''

The Navy noted the crucial flaw in the design in its final report, citing ''the importance of the information being presented by way of the *Vincennes*'s Large Screen [Computer] Displays became apparent . . . It is important to note that altitude cannot be displayed on the Large Screen Display in real-time.''

If altitude readouts had been included on one of the four large computer displays, Captain Rogers himself could have seen the flight was climbing, not descending. As it was, he depended completely on the information he was receiving from his crew members.

As Bruce Coury, a human-factoring engineering expert at the University of Massachusetts said in *Business Week* (no. 3069, September 12, 1988), ''For humans and systems to understand each other, information should appear in the best way possible. That doesn't appear to be the case with the *Vincennes*.''

The confusing manner in which the critical information was given to the crew helped them succumb to what psychologists term ''scenario fulfillment.'' Under the stress of being in combat for the first time, crewmen reported what they thought they ought to be seeing, rather than what they were actually seeing. As talk of an F-14 approaching the ship went across the intercoms in the Combat Information Center, crewmen came to believe the ship was coming under air attack, and may have unconsciously distorted the stream of data they were witnessing to fit the scenario they thought was taking place.

Connected to the outside world only by the Aegis computers, crewmen in the CIC knew only whatever the computer software was telling them. And the confusing design of Aegis further complicated the task of trying to sort out important information.

Psychologists point out how easily expectations can alter perceptions of reality. Even in a non-stressed environment, subjects shown a playing card with a black ace of diamonds will often report seeing an ace of spades or a traditional red ace of diamonds. And they do so with the utmost in confidence and consistency.

Prof. Richard Nisbett of the University of Michigan psychology department has studied this incredible phenomenon. "When you watch subjects in experiments like this, knowing the trick, you can hardly believe the subjects' behavior. While they are looking directly at the object, they insist on something that is being contradicted by their vision. Expectations can color what people think they see and what they remember to a far greater degree than is recognized by common sense."

The crew of the *Vincennes* was in combat for the first time, and tensions were understandably high. From the time the Airbus was first detected to the time it was shot down, Rogers had only seven minutes in which to make a life-or-death decision.

Crewmen gave wildly differing reports of the Airbus's altitude to Navy investigators. Aegis offers two means of determining a plane's altitude. The powerful AN/SPY-1A radar on board the *Vincennes* can determine the altitude of any target within its range. In addition, the UPX-29 IFF receiver could read the altitude directly off the transmissions of the Airbus's transponder.

Although both clearly indicated the Iran Air flight was near a standard cruising altitude of 12,500 feet when the missiles were fired, crewmen recalled seeing altitude readouts on their displays of anywhere between 4,500 and 10,000 feet, a difference of up to 5,500 feet. No one recalled seeing the correct altitude. One crewman told investigators the target was only at 7,800 feet, and he was certain of what he thinks he saw, saying "That [reading] I haven't been able to get out of my mind."

The *Vincennes* crewmen in the CIC were clearly rattled by their first experience in combat. The computer tapes from the ship left behind a remarkable telltale of exactly how the crew had reacted. Wisconsin Congressman Les Aspin says, "We know that one officer, who was prompted by the computer to 'select weapon system' as the countdown to the destruction of the Airbus began, hit the wrong buttons five times before he realized that he was supposed to select a weapon. And we also know that another member of the *Vincennes* crew was so agitated that he got ahead of the firing sequence and pushed another button twenty-three times before it was an appropriate part of the procedure . . . [this is] the norm when inexperienced humans face a sudden stressful encounter.''

Aegis designers, in their search for state-of-the-art efficiency, had neglected to consider how its all-too-human operators would interpret data in a typically distracted combat setting.

Critics angrily charged that the Navy's tests had amounted to having the operators do little more than play video games. The system, and its operators, had never been tested in anything close to simulating the heat and stress of battle. When the government looked into Aegis testing, they found Aegis's true capabilities in an actual combat setting were still a question mark. ''The absence of stress,'' noted a GAO report, ''biased the results in favor of the Aegis, and left actual performance in a more realistic and stressful environment unknown.''

The *Vincennes* and its crew had been subjected to several drills simulating attacks. Some involved having up to twenty-five planes attacking the *Vincennes* at the same time. Still, Comdr. Paul Rinn of the USS *Samuel B. Roberts* readily concedes, ''the stress, of course, we confronted at that time, was certainly not the same as the stress in combat.''

In addition, the crews had not been extensively trained to separate hostile from non-hostile aircraft. Defense Department researcher Dr. Richard Pew, who studies Navy training methods, notes, ''Most [Aegis] training exercises focus on practicing procedures associated with identification and successfully firing on threatening targets. Training that presents targets that could turn out to be threatening but are not is rarely introduced, but it cuts to the heart of the kind of military operations that are most common in a peacetime war.''

The Navy blamed the incident on human error. It points out the computer tapes show that Aegis performed flawlessly. Admiral Fogarty, the chief investigator, concluded, ''The Aegis combat system's performance was excellent—it functioned as designed.'' But, although mechanically it worked perfectly, Aegis failed miserably in its mission. The complicated Aegis software was simply churning out more data than the crew could readily digest. Given the stress of the moment, the crew had simply become overloaded.

Aegis was too complicated for its own good. "Never before in battle," military historian John Keegan told *Newsweek* (vol. 112, August 15, 1988), "has information come in to a commander with the speed and volume that comes into the Combat Information Center of a modern warship." As one retired rear admiral told the magazine, "If you can't interpret what Aegis is telling you when you're under stress, what use is it?"

Operators aboard the *Vincennes*'s escort ship, the frigate USS *Sides*, using a simpler, older radar system, correctly noted the Airbus was at an altitude of 12,500 feet and was climbing, not descending, while the crewmen aboard the *Vincennes* reported the plane was descending out of 9,000 feet. The *Sides* also never received the mysterious Mode 2 military transponder signal. Unfortunately, the information was never passed on to the *Vincennes*.

The stress of the Strait of Hormuz incident showed that Aegis was at once both too complicated and too simple: feeding the crew more data than it needed, while omitting information that would have simply told whether the plane was climbing or descending.

All Aegis systems on the cruisers have since been changed to correct their critical design deficiency. It was a simple software redesign. Information on the altitude of airplanes is now prominently displayed on the large tactical display screens.

As Carlucci noted, solving Aegis's problem in its early design stages could have been as simple as adding an up-or-down arrow that would have clearly indicated whether the plane was climbing or descending.

Ironically, Rogers himself had noted the potential for trouble, in a press conference shortly before the *Vincennes* left San Diego for the Persian Gulf. "There are risks out there," he noted, "and maybe the worst one is the possibility of someone making a mistake. In a confined area like the gulf, and with the weapons at hand, the potential for disaster is incredible." Retired Adm. Eugene Carroll said afterwards in *Maclean's* (vol. 101, August 29, 1988), that with the reliance on computers, "this sort of event was inevitable."

Some in the Navy and at RCA, after reviewing the tapes of the incident, blamed the crew of *Vincennes* for being over-zealous. They say the men aboard *Vincennes* were scrapping for a fight and a chance to prove the worthiness of the much-maligned Aegis system. Even prior to the Airbus incident, the aggressive nature in which *Vincennes* had dealt with other Iranian ships and planes had been the topic of many discussions in Navy wardrooms in the gulf. Crewmen on other ships had nicknamed the *Vincennes*, ''Robo-Cruiser,'' for its eagerness in challenging hostile ships and planes that approached it.

Many critics argue the Navy's report cast too much blame on the crew, and that the Navy was reluctant to admit the other reason why its ultra-sophisticated, $600 million system had failed in the crunch. Aegis was simply the wrong system in the wrong place; Aegis may function well in the open ocean, protecting wide paths of battle groups and convoys, but not in the twenty-five-mile-wide Strait of Hormuz. The Navy admitted the sophisticated Aegis had been thrust into a battle situation for which it had not been designed. ''The current tools used by the U.S. Navy for differentiating between friendly and hostile unknown aircraft,'' said Admiral Fogarty, ''were designed primarily for the open ocean environment.''

By being placed in a narrow body of water where decisions had to be made in short time intervals, the Aegis system lost its primary advantage: its uncanny ability to detect threats while they are still hundreds of miles from the ship. The tactics of the Aegis defense often call for air support from Navy carriers to assist in dealing with incoming missiles and aircraft.

By building the system for combat in the open sea, Pentagon designers had failed to anticipate the scenario of how to use Aegis in a crowded body of water. Although Flight 655 was in constant radio contact with the Bandar-Abbas control tower, the Aegis control center was not designed to monitor common radio communications between air traffic controllers and commercial flights. Aegis was not meant for the half-war, half-peace atmosphere of the gulf.

Aegis was never designed to have to differentiate between hostile and non-hostile aircraft. "It should be appreciated," Admiral Crowe notes, "that Aegis was never advertised as being capable of identifying the type of aircraft being tracked."

Defense Secretary Carlucci brushed aside suggestions that Aegis had been improperly designed. Despite acknowledging its shortcomings, Carlucci said, "I'm not indicating it wasn't designed correctly. As you go through experience with any weapons system you improve the design."

But in reality, Aegis had been designed without consideration of what type of real-world situations it might be called on to encounter. And it never was subjected to the type of testing that could have revealed its fatal flaw.

The Pentagon defended Aegis, saying the super-sophisticated system worked as designed. But as one disgusted congressional staffer put it, "They [the Navy] will do whatever they can to save this piece of hardware. They spent a half billion dollars for the computer system on this boat. And if you're going to spend a half billion dollars, the doggone thing better know what the fuck is up there."

Chapter Ten

OF CRIME AND PUNISHMENT

SHEILA Stossier bid good-bye to the last passenger to deplane from Eastern Flight 91. Sheila sighed with relief. Although she was still adjusting to the rigors of her new job, she was loving every minute of it. Being a flight attendant offered a chance to meet people, and to fulfill her love of travel.

It was, she wistfully reflected as she collected her travel bag, a hectic time. She had started with Eastern barely three weeks earlier. And now, she had just completed another international trip, a weekday jaunt to Cancun, Mexico. Fun in the sun was nice, she thought, but what she really wanted was to be home in New York. Her daughter Danielle, barely two years old, had been sick with asthma. Sheila had been reluctant to take the trip, even though her mother-in-law, she knew, would be taking fine care of her. Danielle had just been released from the hospital that week, and Sheila was anxious to see how she was.

As she and her crewmates left the plane, it was just a short walk from Gate 5 down Concourse *C* of New Orleans International Airport to the Customs station. As was customary, the crew members handed over their passports and crew forms. A Customs agent entered Sheila's name into the computer. The computer conducted a routine search through files held by the FBI's National Crime Information Center computer in Washington, which tracks missing criminals.

The NCIC is one of the largest computer criminal history files in the world. It holds records on some twenty million Americans. Nearly one million requests for information are made to the NCIC every day. It has proven invaluable in tracking down and arresting criminals as they wander from state to state.

But its software is designed in a manner that leaves it vulnerable to misuse and human error. The failure to design crime computer software to minimize human error can mean trouble for the hundreds of thousands of people who cross paths with such software every day. They are not readily designed to allow correction of inaccurate information. Conversely, they are not designed to send information that could prevent misidentifications and subsequent wrongful arrests.

Crime computer software programs, since they are designed to screen people, often work contrary to the principles of justice. Instead of beginning with clues from the crime, and building suspects, crime computers turn the investigative process backwards. People are often checked out, even when there is no suspicion they are guilty of any crime.

Police in patrol cars can randomly check license plate numbers to see if the cars are stolen. People being hired for sensitive government positions are screened for past criminal activity. And people entering the country can be checked as they enter Customs.

SHEILA waited patiently for the Customs agent to clear her. Approval usually took a matter of seconds. Sheila had already passed through Customs in New Orleans seven times that month.

When the reply came back, the Customs agent had a quizzical

expression on her face. "Have you ever used the name Shirley?" she asked Sheila.

Sheila was puzzled. "No," she replied, wondering what the questioning was all about. "My married name is Sheila Jackson Stossier. Jackson is my maiden name. I've been married since 1980. I use both names."

The agent left, leaving Sheila and the crew mystified. She returned with a supervisor. They asked Sheila to step aside, and continued processing the rest of the crew. Finally, the agent told Sheila the news: she would not be continuing on with the crew to New York.

"What's the problem?" Sheila asked.

"We have an arrest warrant. You're wanted for a parole violation, and you'll be sent back to Texas for prosecution."

Sheila laughed. They must be joking, she thought.

Sheila was always, as she described herself, "a good girl," never one to get into trouble. After attending the Texas Southern University School of Journalism, she entered the Marine Corps, where she served for eight years. After leaving the military with an honorable discharge and a rank of sergeant, she turned her attention to the airlines, and promptly became the first female Vietnam-era veteran hired as a flight attendant by Eastern. She had never in her life been in trouble with the law.

The Customs agents, Sheila noticed, were not laughing.

"There must have been a mistake," she said. "I've never been arrested, and I haven't been in Texas in ten years." Sheila showed the Customs agents the stamps on her passport that indicated she had been cleared through Customs several times before. The agent merely turned to her supervisor and snorted, "I wonder why they never stopped her before?" The agents looked at her identification card from the U.S. Marine Corps Reserves which had a fingerprint on it that easily could have been checked against the records in the Houston Police Department. They chose not to. Airport security police had arrived and announced they were taking her into custody.

"Tell Eastern to get me out of here!" Sheila yelled to her crewmates as the police led her away to the main terminal. As she walked through the terminal, Sheila thought to herself, stay calm. This has got to be some kind of a mistake. Sheila Jackson, after all, she thought, is a common name. Once they figure out I'm not the person they're looking for, this will all get straightened out, and I'll be on my way.

Sheila was led to a small room where she was searched. As evening approached, police from Kenner, Louisiana arrived. Sheila was handcuffed and taken to the jail. When she arrived at the police station, Sheila tried to explain that she was not the person they were looking for. An officer told her to shut up and sit down. She asked the officer to verify her identification with Eastern and the Houston Police Department who had issued the warrant. Said Sheila, "They refused. They outright refused."

She was then fingerprinted and photographed. For the first time in her life, Sheila Jackson Stossier now had an arrest record. Allowed to make the customary one phone call, Sheila decided to try and reach a friend of hers who worked for the FBI Special Service Unit in New York. Surely he can straighten this out, she thought. Fortunately, she got through. Special Agent John Pritchard told the officer it was impossible that Sheila was the woman they were looking for. Pritchard told the officer that a simple check of the description and fingerprints of the suspect against Sheila's identification would easily reveal the discrepancies between the two. When Sheila was booked, she had in her possession her passport, her Eastern ID card, a driver's license, a Veterans ID, a social security card, and an Eastern Airlines medical insurance identification card.

As Sheila would later learn, the real "Shirley Jackson" police were searching for was two inches taller, twenty-five pounds heavier, had a different birth date, and used a different social security number. Despite the physical differences, officers said Sheila fit within the "range" of the description of the suspect.

Twelve years earlier, when Sheila moved from Texas, she barely paid notice to the fact her student ID card was missing. When "Shirley" was arrested in Houston in 1979, she apparently gave police Sheila's name and old address, which was recorded into police records, and forwarded to the FBI's NCIC computers in Washington.

An officer came by the cell and asked Sheila for her social security number, phone number, address, and employment. What a relief, Sheila thought. A simple check of the information will show I'm not the person they think I am. At the time, Sheila didn't know the information was simply being collected to be put into her arrest file.

Later that night, Sheila was transferred to a holding cell at the Jefferson Parish County jail. By this time, Sheila was frightened and exhausted. Because officers expected her to be extradited to Texas shortly, she was put in a holding cell with no bed. Sheila asked if the pictures and fingerprints of the suspect had come in from Texas to clear her. The officer delivered the stunning news: They had, and she had better settle down for a long stay, because she was the one they were looking for.

As it turned out, at no point did police ever compare fingerprints or photographs. Sheila curled up on the cold concrete floor and tried to sleep.

Unknown to her, Sheila's friends were trying to track her down and aid in obtaining her release. Glenda Moss, a New York union representative for Eastern's flight attendants, managed to reach Louisiana authorities, and pleaded that Sheila had been employed by Eastern Airlines since October 9, and a thorough screening had revealed no criminal history. Eastern's computers, she pointed out, could verify where Sheila had been on any given day, and she offered to do whatever she could to prove Sheila was not a parole violator. Moss's pleas were ignored, and she was told nothing could be done for Sheila.

At 4:00 A.M., Sheila was transferred yet again, this time to the Jefferson Parish Correctional Center. Desperate, Sheila called out to a corrections officer, Vanessa Shropshire, who was passing by the cell. "I'm not the person you're looking for," she implored.

"Are you telling the truth?" Shropshire asked. All criminals claim they're innocent.

"I swear to you I am," Sheila said. "I have a sick two-year-old at home. I've got to get out of here so that I can be with her."

Shropshire chose to believe Sheila, and helped her verify her story. Shropshire made simple checks—calling Eastern to find out how long Sheila had been employed there and how long she had lived in New York, information that could have been gotten by any of the Customs officials or arresting officers Sheila had encountered in the last eighteen hours.

Finally, at midnight Saturday night, more than twenty-four hours after it all began, Parish Judge Tom McGee authorized Sheila's release upon the posting of a bond for one thousand dollars. Eastern put up the bond, and Sheila was freed. Although she was now released from jail awaiting trial, there was one final indignity: she could not leave the state.

November 2, five days after her arrest at New Orleans International Airport, a parish judge signed an order allowing her to go home. The return home did not end Sheila's troubles. Although she had never been in trouble with the law, she now had an arrest record, even if the arrest had been made mistakenly. Her married name, which she had given to authorities, was now listed as an "alias." Even after returning to work, Sheila complained, "I've suffered from medical difficulties, stress, and inability to sleep. I've been subjected to rumors and hearsay." The pressure finally forced Sheila to move her family to the Washington, D.C. area.

On the day that the warrant was issued in Houston for the arrest of "Shirley Jackson," Sheila Stossier had passed through Customs in New Orleans. Her passport was stamped with the date, September 13, clearly indicating her whereabouts that day. The fact she was carrying a passport at all should have tipped off authorities. Sheila had been issued her passport May 16, 1980. Persons convicted of felonies and who are on parole cannot be issued passports. "Shirley Jackson" had been arrested January 9, 1979, sixteen months earlier.

The nature of crime computer software is that wrongful arrests are never expunged from the system. If a person is arrested because of computer mix-up, the record may be corrected to show they were later released. But it will still carry the record of the arrest, which could conceivably cause problems when people apply for employment, or if they happen to be stopped on suspicion of other crimes. Innocent people will now carry a cloud of suspicion with them, because their records indicate an arrest, even if it was all due to a mistake.

And then there is always the danger the nightmare could repeat itself. Anyone who checks the FBI's NCIC crime computer will find Sheila Jackson Stossier has an arrest record. As Sheila's attorney, Theodore Mars, noted, "If she passes Customs again and they punch Sheila Jackson, or she goes to apply for a job under Sheila Jackson, the bad Sheila Jackson's record will always come up, and I don't know of any way of having that expunged." In fact, if any government agency has permission to check her NCIC file, it will show she was arrested on an alleged felony.

Worse, if Shirley commits further crimes, and is in fact using Sheila Stossier's old ID card, Sheila could be arrested for it. In a different case in Louisiana where a woman was falsely accused of a crime, a police officer told her, "You had better go out and change your name, because you have too common a name, and the way the computers are run around here, you're just going to keep getting arrested and rearrested."

The irony is that even Sheila's initial arrest at the airport was easily preventable. The Customs agents simply relied too much on the information provided by the computer to tell them they had the right person. As her attorney points out, "It seems in her case, if just a little common sense had been applied, this lady would not have been forced to undergo all she did."

The case is by no means isolated. Rep. John Conyers, Jr., (D. Mich.) says "I have been told there are hundreds, probably thousands" of people wrongly arrested because of mix-ups in the NCIC computer system.

Atty. William Quigley found hundreds of wrongful arrests because of computer errors in New Orleans alone. "I've received inquiries from attorneys all over the country who had similar problems. It is not an isolated thing."

The problem is inaccurate, outdated, or misused information. Quigley defended a woman named Shirley Jones, who was detained by New Orleans police. When applying for foster parenthood, Jones agreed to a routine background computer search. The very next day, two police officers knocked on her door. They asked to speak with a Vera Davis. Puzzled, Jones said nobody by that name lived there. The officers demanded Shirley identify herself. She did, showing identification as Shirley Jones.

The police then informed Shirley she was under arrest for forgery and theft charges. Shirley protested that she was innocent, that it was a mistake. The officers said they were just doing their job, and took her to the downtown central lockup, leaving her children at home. Shirley was booked, fingerprinted, photographed, and her name entered into the computer.

Twenty hours passed before Shirley appeared before a magistrate. Luckily, she appeared before a black magistrate who sympathized with her pleadings, and released her.

Attorneys soon discovered Shirley had been arrested merely because her name, Shirley Jones, matched up with one of a dozen different aliases used by a suspect named Vera Davis. The two women had little in common, other than that they were both black. The two women in question had different birth dates, were six inches different in height, and seventy pounds different in weight.

Even after these differences were pointed out, the sheriff's department insisted they had arrested the person they were looking for, and continued to press charges. A judge, upon hearing the evidence, quickly dropped the charges.

Quigley, who serves as general counsel for the Louisiana branch of the American Civil Liberties Union, began an investigation into the New Orleans police computer files, and uncovered a huge mass of problems.

Although the New Orleans Police Department maintained a full-time staff of qualified, trained professionals who worked on the computer, Quigley found a tangled hodge-podge of at least twenty other agencies that had access to the computer and the ability to alter or delete records in their own jurisdiction. Traffic courts, municipal courts, parish deputies, were all able to use the system without any specialized training.

The system was also awash in old information. Arrest warrants several years old were found still floating in the computer's memory banks. In many such cases, the suspects authorities sought had already been arrested. But unfortunately, the software did not require users to enter in an identification number when logging in. Because of that, it was impossible to track who had entered the information, and who had later gone back into the file to update it. Without being able to trace who had made the mistake in the beginning, tracking down the errant user to have them correct the error was next to impossible.

People going in to renew their driver's licenses or vehicle identification stickers were routinely stopped when the computer showed they had an outstanding ticket, had not paid their taxes, or had failed to appear in court. People began carrying their receipts with them every time they went into motor vehicle division offices for new licenses because they were used to having to explain why their name kept coming up on the computer.

The problems got worse. Because of its frequent inaccuracy, police began questioning the data in the files. If a suspect was anywhere close to even vaguely matching the description of a wanted suspect, Quigley says, the police would simply shrug it off and say, "Well, you know how the computer is. There are problems here and there. I'll just go ahead and take a leap and say that's the person, and we'll go ahead and arrest them."

Quigley once visited a person who was arrested even though his birthdate did not match up with the real suspect's. "The deputy said, 'Well, you know, it's close to the same month, and it's only one year off; then that's pretty close.' " Quigley was appalled. "He wasn't joking, and he wasn't trying to be smart. That was the way he viewed the accuracy of the information in the computer."

One man was arrested because he shared a similar name with a suspect. The problem was then compounded by a second computer error. When he was booked, his last name was incorrectly entered, through a typing error, into the computer as "Brussell" instead of "Russell." The man languished in jail for forty days, because the computer, searching in vain for the correct name, never put the case on the calendar, and he was never called to appear in court. In addition, Russell was slightly retarded, and never inquired about his status while in jail. Worried family members had contacted police, but they could not locate him because the computer said no such person was in jail.

A cellmate was released and told family members of Russell's predicament. But even then his release was plagued with problems. Jail officials refused to acknowledge he was there, even though family members knew precisely which tier his cell was on. Jailers had no record of his being jailed there, and insisted their computers were correct. It took two days to straighten out matters and obtain Russell's release.

Even an attorney representing the City of New Orleans admits there are problems with information being incorrectly entered into the computer. "Someone pays a ticket," Tom Anzelmo says, "it doesn't make it on the computer that the ticket was paid, and then a warrant is issued for their arrest. We've had maybe six of those."

Anzelmo says that as an attorney, he has defended the city in more than one hundred lawsuits involving wrongful arrest because of computer problems. Many plaintiffs have been successful in recovering damages for loss of their constitutional rights. But as Quigley points out, "an overwhelming number of people don't even bring suits. They just have to lump it."

Similar problems can be found in local crime computers throughout the country. Assoc. Prof. Diana Gordon of City University of New York became involved in a project called Justice Watch that investigated computer problems in the criminal justice system, and found "the wrong warrant problem is an illustration of this kind of problem. I should say the problem exists very substantially on the local level in other areas besides New Orleans."

The problem is perhaps greatest in the FBI's National Crime

Information Center computer. Begun in 1967 with a mere three hundred thousand files, the NCIC was simply a means for states and the FBI to exchange criminal records. But as the files grew, new data base search software, coupled with the proliferation of NCIC terminals, allowed more and more law enforcement personnel around the country an opportunity to screen people in hopes of finding criminals. Quigley notes, "Everybody has it now. Every district attorney's office, even the smallest little traffic court has a computer hookup with the major metropolitan crime information computer." More than sixty-five thousand different agencies around the country have access into the NCIC files. As a congressional advisory panel noted, "Computer systems can be used as an electronic dragnet, sifting through huge volumes of information. The wider the net, and the more frequently it is cast, the greater the threat to the civil liberties of those caught in its strands."

As it is currently designed, the NCIC software provides only enough information to match up potential suspects. It is not comprehensive enough to include things such as fingerprints or photographs that could prevent misidentification and wrongful arrests.

Just before Christmas of 1988, Customs agents at Los Angeles International Airport were routinely running the names of passengers deplaning a TWA flight from London. The NCIC computer came back with a "hit." Richard Lawrence Sklar turned out to be wanted for a real estate scam in Arizona. The information provided by the NCIC was dead-on accurate: the height, weight, eyes, and hair color all matched. Sklar was handcuffed and taken into custody. The only problem was, he was the wrong man.

Sklar had committed no crime. It turned out he was a fifty-seven-year-old political science professor at UCLA. Over the next two days he would find himself handcuffed to gang members, strip-searched, and moved from jail to jail. Someone in Arizona had assumed Sklar's identity, possibly through a stolen ID, and was using his name to commit a series of financial misdealings. The Sklar case points up the flaw in the software. Janlori Goldman of the ACLU sums it up: "The FBI conceded after this event that this is the kind of NCIC error that is currently unavoidable.'' The software provides just enough information to arrest, but often not enough information to prevent wrongful arrests.

To further add insult to injury, the NCIC computer software is designed in such a way in that it provides almost no opportunity to correct or remove errors. Professor Sklar ended up being arrested a total of three times on the same misinformation.

A Michigan man was arrested five separate times as a result of an error in his NCIC file. Terry Rogan lost his wallet while on a camping trip. Someone picked it up and began using his identification while committing a series of murders and robberies. As Rogan's camping trip progressed, he was arrested four times in four different states. The fifth time, Los Angeles police travelled all the way to Saginaw, Michigan to arrest him. Even after releasing him, the LAPD refused to agree to correct the information in his file.

Rogan finally went in person to the FBI in Saginaw and angrily told them he was tired of being arrested at gunpoint. The FBI agent told him the FBI couldn't do anything about it, and that he ought to contact his congressman.

As William Quigley points out, ''There are individuals being charged with crimes where there are basic similarities in names, ages, and weights that are resulting in the wrong person being apprehended. It usually results in someone's rights being detrimentally affected because they are stopped, they are detained, and they are questioned.''

Occasionally it reaches the point of absurdity. In one case in Los Angeles, a black man was arrested because he shared the same name with a white man's warrant.

Michigan Congressman John Conyers, Jr., throws up his hands and admits, ''I'm getting an increasing sense of frustration. We're supposed to be solving this with the FBI, and right now I haven't got a solution in the world.''

Law enforcement officials in California and New York say that at various times, as many as twenty percent of the warrants in the NCIC system are stale or invalid or simply have not been removed from the system. As Professor Gordon says, "any file which relies on local standards will have a lot of errors in it."

The FBI has neither the time nor the resources to go through and verify the accuracy of its millions of records. The agency does have periodic data quality audits, but it is physically impossible for them to check and recheck every record every time it is updated or expires.

One common problem is the failure of courts to enter the final dispositions of cases into criminal records. Two California residents applied for licenses to become substitute school teachers. They were both denied approval because computer records showed they had been arrested. In both cases, the charges had immediately been dismissed. But the courts had failed to send the proper forms to the state records department documenting the dismissals. Accordingly, the dismissals were never entered into the computers.

Professor Gordon notes, "For many types of offenses and in many places, well over half of the cases are either dismissals or acquittals instead of convictions. But that often doesn't get put into the record." California conducted a survey of its computer files and found that more than one hundred thousand files had unrecorded dispositions.

Another problem is that the software used in crime computers often allows unauthorized access to the information held in its files. The American Civil Liberties Union discovered anybody in the New Orleans area could find out what a person was wanted for, what their rapsheet looked like, and what their criminal history was, without either the permission of the person or the knowledge of any applicable law enforcement agencies. Merely by knowing someone who worked in the sheriff's department or a traffic court, the information could be easily summoned up without anyone ever knowing there had been an unauthorized access to the New Orleans police files. Cases were found where employers had used the crime computers to run background checks on potential employees without their knowledge, a violation of data privacy laws.

In one instance, an employee showed up for his first day at a new job, and was greeted by sheriff's deputies who informed the startled employee, "You're under arrest, because you're wanted for armed robbery." Despite his protestations, the young man spent twelve hours in jail before police realized they had the right name but the wrong person.

Gordon says examples have been found where NCIC rapsheets were sold to private investigators. In two instances, Defense Department investigators attempted to gain access without authorization into state criminal history records. "The portability of these kinds of records," she says, "poses a great temptation to use them for purposes for which they were not originally intended. There are potential problems of abuse of computerized records for the suppression of political dissent or for blackmail."

Professor Gordon says such records could be misused in political campaigns and espionage activities. "Even if there were no mistakes in this kind of file, and even if there were no abuses of it, I am still very concerned about the general trend towards unregulated expansion of computerized files which may influence criminal investigations and, at some future date, employment investigations. There is a seemingly unending stream of expansion of files at all governmental levels. Overall regulations of these kinds of records is grossly inadequate."

Criminal data bases are growing quickly. Cities and counties are now developing their own computer files. And the network grows as local files are shared with other state and local computer systems, and ultimately the NCIC in Washington. The possibility for other municipalities updating or deleting files without coordination grows. "There's nothing that removes data on a regular basis," Professor Gordon says. "There's just continuous addition of data. The record may be corrected later, but you don't know all the states to which the incorrect record has been sent. It's an immense area of absolutely unregulated police monitoring. There are more and more data bases, and more and more people who have access to them. It's a huge, huge subject."

New Orleans, for one, has made great strides in improving its police computer system: anyone with access to view or update files must now be trained and certified, with periodic updates. The city has made important changes to its software. All users must log in with individual identification numbers before entering into criminal files, leaving behind a record of who has looked at particular files. The files themselves are regularly cleaned out and updated every six months.

The ACLU's William Quigley points out, "The NCIC needs to be upgraded and improved just the same way as the New Orleans area. The system can be improved, and people can work cooperatively in a way that does not unduly restrict the legitimate authority of law enforcement to investigate and arrest criminals. The system can be improved to minimize the number of victims. In addition, our experience is that there is not a tremendous price tag on that."

"We understand there are always going to be mistakes," he says, "but it seemed to us that maybe there were mistakes they were repeating and maybe there were mistakes that could have been prevented."

FBI director William Sessions acknowledged before a congressional hearing the need to "improve data quality and enhance privacy and civil and constitutional rights protections." The FBI is attempting to improve the NCIC. Training programs attempt to eliminate errors at the source, the local and state agencies that use the system. Audits are conducted every two years to sweep out old records and check the accuracy of new files. States that show a high number of problems in audits are watched more closely.

Janlori Goldman heads up the ACLU's Privacy Project, which analyzes privacy issues involving various electronic law enforcement equipment. She says, "Data quality is improving, and we understand the FBI also shares this concern, but more needs to be done." Goldman worries that the FBI is moving ahead with plans for wider, more comprehensive computer systems without having first solved the problems of how to design the software to avoid mistakes. "Even though we understand the FBI shares this concern, they continue to move forward with additional expansions and further troubling proposals without first addressing this issue.''

The FBI is studying a proposal for a new computer system called NCIC 2000. It would provide more detailed information, including fingerprints and photographs. The images would be electronically transmitted directly to the officer making the inquiry, even in police cars.

Asst. Director William Bayse, the official in charge of FBI technical services, says, ''the ability to make automated fingerprint [and photograph] comparisons might reduce the incidents of wrongful detention, such as in the Professor Sklar case.''

Bayse points out that NCIC 2000 will contain software that is designed to detect errors and keep old records out of the system. As he says, ''now someone enters data in California, it comes into Washington to find out the data is wrong, and the error goes back to California. We want a richer set of edits so that it never gets onto the network, never gets into the system.''

NCIC 2000 software will also include built-in ''audit trails.'' Users would be required to log on with code numbers that show who accessed particular records, and what updates were made to them. But ultimately, Professor Gordon notes, ''The quality of NCIC information is only as high as the quality of the data put into it from the local sources, and that still seems to be insufficient; deficient. The previous studies by the FBI itself show many stale and inaccurate records. And we have found a lot of acknowledgment that warrants in the record system—local and state and federal—have already been served or discharged or contain errors. And that arrests in criminal history files are often recorded without the disposition which followed. I interviewed law enforcement personnel in seven states who used and managed computerized records. One of my interviewees said to me, 'We're lousy housekeepers,' and this has to be acknowledged, I think, by the designers of this system.''

Despite advances in software technology that will allow the FBI to catch and correct many errors as they enter the system, the software design for NCIC 2000 is still lacking in one critical area. As currently designed, the NCIC 2000 plan does not include a procedure for someone to request a correction of his or her record and get it.

David Nemecek, the FBI's NCIC section chief, admits the agency has no plans to include that feature in its NCIC 2000 software. He claims the current procedure for correcting records is sufficient, saying, "There is a very formal available process that the FBI Identification Division has been offering for a number of years. [It] has been honored and has been effective."

But Rep. Don Edwards (D-Calif.) disagrees. "It is very difficult to correct a record today as requested by an individual." Having suffered through five wrongful arrests, Terry Rogan eventually had to sue the Los Angeles Police Department in order to get his NCIC file corrected.

Professor Richard Sklar, arrested three times as a result of a criminal using his name, simply concedes in the *Washington Post* (vol. 112, February 13, 1989), "Until this person is caught, I am likely to be victimized by another warrant."

The built-in audit trails in NCIC 2000 are designed to cut down on unauthorized access to criminal history files. But no matter how carefully the software is written, users can always find a way to get around built-in security measures. As Professor Gordon found out in her studies, "Investigators abuse local and state systems with considerable ease." They would occasionally use an authorization number or an investigation number to open a file for investigation X, but then make a second inquiry about somebody else they're also interested in for another investigation. As Gordon puts it, "They're very inventive in figuring out how to use the data systems in new ways; sometimes unauthorized."

The FBI is making a concerted effort to make certain its NCIC 2000 software will not prove vulnerable to unauthorized access into its tens of millions of files. But no matter what precautions are taken, a House Judiciary Committee report warned, "A system with thousands of terminals and potentially several hundred thousand users is inherently vulnerable to unauthorized use.''

Then there is the question about what's to be done about the myriad of state and local computer crime systems. Despite the millions of inquiries that are made into NCIC, most crime investigation still relies on state files. As Representative Conyers asks rhetorically, ''What is going to happen? Do we have to bring forty-nine other lawsuits in the several states, and is that going to do any good anyway?''

No one doubts the value of computerized crime history files. It has without question saved many officers' lives by alerting them to potentially dangerous suspects prior to serving traffic stops or search warrants. A simple computer check done from the terminal in a patrol car can clear up any misgivings. A recent survey showed the instant access to the twenty million files in the NCIC resulted in the apprehension of eighty thousand wanted felons and the recovery of $460 million worth of stolen vehicles in 1988 alone. Rep. Don Edwards sums it up this way: ''It is a valuable tool. However, if it is improperly used, it can be a great threat to the privacy rights and civil liberties of Americans.''

As Janlori Goldman of the ACLU wearily notes, ''There are no simple solutions to this data quality problem. But at a minimum, individuals should be able to check and request correction of their own records.'' The problem is partly administrative; law enforcement agencies are understandably reluctant to make it too easy for people to make deletions or corrections to their records. But it is also largely a problem of software design; software for crime computers must be comprehensive enough to provide enough information to avoid misidentification, periodically clear out old records, and provide security safeguards against unauthorized access into files.

Goldman suggests a data quality office, independent of the FBI, be established. Such an office would set strict requirements for criminal history data, and ensure that the FBI and all local and state agencies with access into NCIC adhere to them.

Attorney Thomas Anzelmo, who occasionally represents the City of New Orleans, admits there will always be problems with mis-identification. "There are criminals who use aliases, and that's I guess, an inherent fault. When the alias goes into the system and it's the same name as someone else, then at some point in time, if the two names are compared, you're going to get a positive match." The solution, Anzelmo hopes, is better training that will allow computer users to spot differences in vital information such as birth dates, social security numbers, and physical descriptions.

But ACLU's William Quigley feels that as the criminal justice system relies more and more on computerized information files, problems will grow. "As we store more and more information, there's more and more opportunity for people to make mistakes. I think as searches become more comprehensive, we'll see more of them. And therefore, I think we'll see more problems." Already, the FBI receives ten inquiries every *second* into its NCIC files.

So the battle for a more error-free national crime computer system will go on, and training and updating, and more flexible software will have to be a constantly evolving process. "No one is ever going to make an error-free information system," he says. "But maybe some of them are preventable. The system can be improved. But our computer experts assure us that no matter what we do, there are always going to be new problems."

Chapter Eleven

THE HUMAN ELEMENT

WHY do people and computers have such a tough time getting along? Why is it so difficult and often painful to get computers to do what we want them to? After all, people often fancy that computers are little more than mechanical and less elegant versions of the human brain. In fact, they may in some ways be our superior, being incredibly more efficient both in terms of speed and capacity.

But, in truth, nothing could be further from reality. For there exists a fundamental mismatch between people and computers. Although people certainly created computers, they did not create them in their own image.

And the human element, namely the people who will be using the computers, all too often gets neglected when computers are being designed.

For starters, computers don't even think the same way we do. We use vastly dissimilar processes for dealing with information and solving problems. Computers can process enormous amounts of information. But they do so in a form that is nearly incomprehensible to people. To a computer, everything must be expressed in terms of numbers. Every piece of information must be broken down into numbers that can be digested by the computer's binary brain. Everything, in its very simplest form, must be expressed either as a 0 or 1 so the computer can process it in that form.

John Rollwagen, chairman of Cray Research, heads up the world's foremost manufacturer of supercomputers, machines that possess enormous speed and computing power. But he notes that fundamental differences exist between humans and computers and the manner in which each processes information. "There is a tremendous impedance mismatch between us and computers. We do not think in the same terms. I defy any of us to read three and a half pages of computer output and make any sense of it whatsoever."

Dr. William Wulf is one of the nation's leading experts on software. As Asst. Director of the National Science Foundation, he heads the NSF's Computer and Information Science and Engineering Department. He notes that computers can handle with considerable ease, enormous volumes of data that would befuddle even a nuclear scientist. "The amount of data that results from the kinds of scientific computations that we are talking about is enormous. It is not the sort of thing that human beings can understand by looking at a pile of paper ten feet high." And yet that is just the sort of information a computer can put out without batting an electronic eye.

The enormous sums of information a computer can produce is so large as to be incomprehensible; making it comprehensible to humans, that's where the trouble begins.

As Dr. Karl Winkler of the National Center for Supercomputing Applications at the University of Illinois notes, "What is the bottleneck? Are the humans the bottleneck in the data transformation game, or is it the technology, the machinery around us? We humans are the limiting factor."

Knowledge is power. But information is not necessarily the same thing as knowledge. Computers can throw information at people far faster than they can process it, and in far greater quantities than they need to use. What computers do best, process information, can easily overwhelm people and hinder, rather than help us when making decisions.

A more fundamental problem is that computers simply do not

think by the same methods we do. All computers operate by using algorithms. That is to say, all problems must be solved through the use of a specific set of rules. Everything must be reduced to a series of yes-or-no questions.

There is no intuition or abstract thinking in the mind of the computer. Winkler notes there are some things we do better than computers. "Humans have by far the best pattern recognition activities compared to any system we ever built. We have the ability for abstraction."

As NSF's William Wulf puts it, "we as human beings process and understand information in a particular way."

This puts people and computers on a potential collision course when it comes to decision-making.

Ideally, computers are meant to support or help people making decisions. By providing critical information, the computer helps the person, be he/she a doctor, airline pilot, ship commander, or police officer, make his/her decision.

But the critical question becomes, does a better informed person make better decisions? All too often, the use of computer-developed information causes the person to lose control of his/her decision-making to the computer. Dr. Richard Pew of the American Psychiatric Association has been studying the effects of automation on decision-making for twenty years. He explains, "The use of automation to reduce workload is a mixed blessing. The operators tend to regard the computer's recommendation as more authoritative than is warranted. They tend to rely on the system and take a less active role [themselves]."

It can become an information trap: if the computer says the person is a wanted criminal, he/she must be a criminal. If the display tells a pilot the plane is flying correctly, everything must be okay. If the computer says the incoming plane is hostile and not a jetliner, it must be a threat. "He leans on the system," Pew says of the all-too-typical computer user, "and reduces his reliance on his own judgement."

In contrast with the fact that computers can handle large volumes of information, humans are extremely poor at processing information and using that information to arrive at decisions.

Dr. Paul Slovic of the University of Oregon has studied the science of decision-making since the field was in its infancy in the 1950s. In his studies, Professor Slovic has found that there are several key tasks that must be accomplished for effective decision-making. Clear objectives for the decision must be set. The decision-maker needs to assess what the possible outcomes of each decision are, the probabilities of each outcome, and the consequences of each. Somehow, these judgments must be jumbled into a mental equation that can be used to weigh every possible action.

Unfortunately, though, even under non-stressful conditions, people are inherently deficient in each of these tasks. "We tend to think of only a fraction of all the options that are worth considering," Slovic says. "We have great difficulty judging probabilities or using evidence to infer or diagnose the state of the decision environment."

To compensate, people often tend to rely on the computer to help them make their decision. But they often do so without checking whether the information the computer is giving them is correct or not. "Once [information] is presented on the screen," Slovic says, "the operators tend to treat it as fact." And those "facts," verified or not, are then incorporated into all subsequent decision-making.

The problems get worse when people are under stress, or when the decision is particularly important. Slovic notes, "decision-making becomes very difficult when the stakes are high." Research shows that when placed under pressure, especially life-and-death situations, people greatly simplify their decision-making. Rather than consider a wide range of actions, people will instead tend to focus intently on only one or two key pieces of information to make their decision.

How the computer is designed to handle information then, can have a great influence on what decisions are made. Studies done by the Pentagon's Office of Naval Research show that people usually stop after they encounter one or two possible courses of action.

So, how information is presented can greatly affect the outcome. A study by Stanford researchers showed that when physicians were shown survival rates for cancer patients using two different types of treatment, sixteen percent chose radiation therapy. But when the same information was altered to show death rates, not survival rates, the number rose to fifty percent.

As Professor Slovic notes, "relatively subtle changes in the ways that decision options are described or framed can lead to completely different choices. Slight changes in the way that the judgmental response or decision is elicited can also lead to radically different choices. Information displays do not merely provide information. They interact with the decision task to actually shape or define the resulting judgment or choice."

Whatever information is presented to the person first is the most critical. Office of Naval Research studies show that once a person makes a decision, he or she tends to ignore evidence to the contrary. Worse, information that backs up a decision tends to be counted twice, increasing a person's confidence in his or her initial decision.

Much is still not known about how people make decisions, and how best to present them information that will help the decision-making process. The ONR has been studying this for more than forty years. Each year, the Navy spends more than $12 million analyzing how best to design its computers to facilitate effective decision-making. But as Dr. Steven Zornetzer, Director of ONR's Life Sciences Directorate notes, "These research problems are among the most difficult and complex in the entire research spectrum. Understanding the foundations of human thinking and the many factors which influence human information processing, such as stress and anxiety, is not a trivial problem."

But still, much has been learned about human behavior. And yet, it has not been incorporated into our computers. Professor Slovic complains, "There is considerable knowledge about how decisions go wrong. More than twenty-five years after the documentation of human inadequacies in probabilistic diagnoses, these proposals for decision-making do not appear to have been implemented. Nor do decision-makers appear to have been taught or warned about such deficiencies."

Clearly, computer software needs to be more carefully designed to take into account the deficiencies we, as humans, have. As the American Psychiatric Association's Dr. Pew says, "We need to get human factors analysis introduced at the time the system architectures are being developed. We need to understand how to integrate human operator performance with system performance. We need to get that in at the stage of conceptual design." Unfortunately, Pew notes, the human element is often not considered until the very end stages of development. And by that time, most of the human elements consist of physical things such as how easy are the controls to reach, and how high is the display screen.

Prof. Richard Helmreich of the psychology department at the University of Texas says how to present information in the most effective and timely manner is "a much more serious problem than equipment failure or lack of technical competence." By means of an example, he points out that "in the USS *Vincennes* incident, the highly sophisticated equipment performed flawlessly while breakdowns occurred in the machine-to-commander link."

In particular, more research is urgently needed on how computers can affect decision-making in groups. NASA studies have shown that seventy percent of commercial airline crashes are not the result of technical errors but, instead, as Helmreich notes, are due to "failures of crew performance. These are much more frequent than failures in equipment and much more frequent than lack of technical competence. It is the interpersonal failure that seems to lead to accidents. Indeed, while we have a relatively good knowledge of individual-to-machine interface in automated environments, we have much less familiarity with the group-computer interface, especially under stress, in such environments as an Aegis-equipped cruiser or an automated air transport."

But equally important, humans have to stop relying on computers to provide all the answers in important situations. Computers should assist human decision-making, not replace it. All too often, though, there is an abdication of decision-making responsibility by people who, sometimes subconsciously, let the computer take over, or at least direct the decision.

Psychologists often refer to the "Law of the Hammer": when you give a child a hammer, everything becomes a nail. People similarly rely on the computer to help them make their decision, regardless of whether they feel uneasy about the decision or the accuracy of the information the computer is giving them. Otherwise, what's the computer for?

People must understand they are the ultimate backup to the computer. And they have the responsibility to make sure their decision is correct, regardless of what the computer is telling them.

THE incompatibility of people and computers extends even into the physical design of computers. Modern computer workstations and personal computers are causing physical problems for many of those who have to deal with them every day.

The problem is two-fold: modern computer workstations and personal computers have not been adequately designed to take into account the physical needs of the people using them, and likewise, offices have not been redesigned to accommodate the arrival of the newest piece of office equipment.

The use of computers has skyrocketed in recent years. The number of Americans who work at computer terminals is now 28 million, compared with only 675,000 in 1976. That increase in use has corresponded with another, more alarming statistic: the number of workers suffering from what are known as repetitive strain injuries is also skyrocketing.

Such injuries range from general wrist pain to tendonitis to carpal tunnel syndrome, a painful inflammation of the carpal ligament, which runs across each wrist. Workers suffering from these myriad of ailments often complain of stiffness or shooting pains in the wrist, or numbness in the fingers. In fact, in its most severe form, carpal tunnel syndrome can cause the gradual cutoff of the median nerve, resulting in the gradual loss of use of the thumb and first three fingers. Although problems such as tendonitis can be treated by rest and anti-inflammatory drugs such as aspirin, carpal tunnel syndrome in an estimated five percent of cases requires surgery in order to prevent loss of use of the hand.

Repetitive strain injuries, or RSIs, now account for nearly half of all workplace injuries and illnesses, up from only eighteen percent in 1981. It is the nation's most common cause of workplace injury. In 1989, 147,000 workers in the United States suffered repetitive strain injuries on the job, an increase of 32,000 from just the year before. Although such injuries are also common among meat packers, textile workers, and pianists, the biggest increase recorded by the Bureau of Labor Statistics was by those working at computer terminals.

In the busy newsroom of the Los Angeles Times, more than two hundred of the newspaper's eleven hundred reporters and copy editors have sought medical care for RSIs. Such injuries are becoming increasingly costly for businesses. Susan Burt studies RSI for the National Institute for Occupational Safety and Health. "Some people, it comes and goes very quickly," she says. "But with other people it gets very serious and keeps them out of work for years."

The problem is that computer terminals were not designed with human hands in mind. Skilled typists can make their hands fly across their keyboards at a blistering pace of more than twelve thousand keystrokes each hour. Such rapid, repetitive motions can easily strain the tendons and joints of the hands and wrist. In the days of manual and electric typewriters, typists encountered short "microbreaks" in their repetitive work: pausing to insert a fresh sheet of paper, waiting for the carriage to move to a new line, correcting an error. Even these brief pauses provided enough of a break to relieve tension in hands and wrists. Now, hands and wrists never have to leave the keyboard.

After several reporters at the Long Island newspaper *Newsday* and other news organizations developed RSIs, the various media outlets filed several lawsuits against Atex, the manufacturer of one of the most widely used computerized newsroom systems. The suit charges that the design of the workstations contributed to the repetitive strain injuries suffered by the workers.

But fault also lies at the other end. Office and other workplaces have been slow to adapt work spaces for computer terminals. Workstations are often haphazardly set up on desks, benches, or other work areas designed for other functions.

Having just the right setting for a computer workstation is not a matter to be considered lightly. As Susan Burt of NIOSH notes, "There's been an explosion in the number of people using personal computers, and we've brought them into our offices, and now we're adjusting to them. When you try to set your PC on a desk that was meant for another purpose you run into problems. It requires setting up our offices and desks in different ways."

Experts on ergonomics, or the study of human comfort and efficiency, say even simple changes can greatly reduce problems. Chairs should offer good back support, and have easily adjustable height. The height should be adjusted so that the user sits with his/her feet flat on the floor, elbows bent ninety degrees, with forearms parallel to the floor. Lower legs should run vertically to the floor. The table or keyboard stand should also be adjustable, so that the wrists are level and neither drooping nor raised. A moveable keyboard should be used so that the keyboard can be slid forward or back until the fingers just reach the keys comfortably. Video display screens should be placed at eye level or just below. The line of sight to the screen should be ten to twenty degrees below horizontal.

A variety of helpful devices are also on the market. Wrist support pads run along the lower edge of the keyboard and help support the wrist. Document holders should be used to suspend papers close to the screen and at the same level. Burt notes, "All of the little gadgets that you see: document holders, foot rests, wrist rests, I think all of these things are a good idea that are at least going in the right direction towards supporting someone in the direction they need to be in to work on a computer."

Still, she notes, "There is no perfect posture. I don't think that anyone can sit in an office for eight hours a day and be comfortable." To that end, rest breaks are essential. Frequent, shorter breaks are preferable to prolonged breaks. At a minimum, workers susceptible to RSI should stretch their arms, hands, and wrists for a few minutes every hour.

Preventative measures such as these can help protect against these ailments that some refer to as "computeritis"—yet another sign of the frequent mismatch between computers and people.

MUCH has been said in this book about software errors. Software errors are clearly a devastating threat to human safety, corporate profits, and government efficiency. But many software errors result from a lack of incorporating human factors into software design.

As we've seen, software errors generally develop when a program, be it a business program or one that controls an airplane, encounters a situation that the programmer did not adequately anticipate when he/she developed the program. When the program then encounters an unexpected situation, it either crashes and causes the computer to shut down, or else produces an erroneous result.

The problems often arise because the person using the system (the pilot, the doctor, etc.) is not the same person who wrote it. To write a good program, a software engineer has to fully understand what the program is to do, whether it's fly a plane or run a nuclear power plant.

Jim Webber is a consultant for several Fortune 500 companies on new software technologies. His Omicron group is a consortium of some of the largest computer users in the Northeast, including companies involved in everything from insurance to pharmaceuticals. He notes the frustration in establishing effective communications between those who use software and those charged with the task of figuring out how to write it. "It's a huge problem," he laments. "Users don't have very good ways of communicating what it is they want to do and need to do. And computer analysts don't have very good ways to pull that out of users."

Many times, he notes, the people who want new software programs are simply not inclined to take time to explain in great detail their jobs and what functions they want their software to perform.

"People are busy doing their jobs. They don't want to sit and talk about what it is they do. Instead, they just want to get right to it."

The result, invariably, is software that is late, more expensive than planned, and which does not work as intended.

Software developers must fully understand what the user does, and what the software is supposed to do. The greater that understanding, the better the software that results. A House science subcommittee concluded, "Poor communications between user and developer explains most software flaws."

Former university professor and computer industry consultant Dr. Jacob Schwartz notes, "The reason that a spreadsheet program like Lotus 1-2-3, to take a particular example, is so effective is that it captures an accountant's way of thinking."

Similarly, programs designed to be used in engineering or biology should mimic the thought processes of engineers and biologists. "One has to," he says, "think deeply and carefully about the area, forming just the right mental model of what an accountant does, or of what an engineer or biologist does when setting up the kind of computation for which the program will be used."

Unfortunately, such careful development can be slow. "It takes thinking, it takes false starts, it takes time. One has to find just the right model; that is not easy. Finding the best, rather than the second-best approach has sometimes taken years."

But in the race to keep up with the demands of ever-faster computers and more uses for them, such time is often not available. In an era when, increasingly, time is money, the pressures to rush development of new software are omnipresent.

Patience is one of the keys to proper development of correct software. This key applies to both software developers and users. Barry Boehm, chief scientist at TRW Corporation's Defense Systems Group, estimates that fully sixty percent of software problems are design errors. Sufficient time and money must be set aside early in the lifespan of a software project for proper design from the beginning, as well as for unexpected delays and problems.

As the House Committee on Science, Space, and Technology

notes, "The consensus among software professionals is that the most effective method for reducing problems in software is to allow sufficient time for design." There is no other single solution to the software crisis that promises as dramatic results.

Learning how to effectively merge both people and computers will be the key to solving the software crisis. As David Nagel, a researcher at Apple Computers, urges, "Finally, and perhaps most difficult of all, I think we are going to have to continue to work on the problem of the human interface with these new and complex computer architectures."

The answer lies in a better marriage between people and their computers.

Chapter Twelve

NO MIRACLES

"SOFTWARE is a problem of major importance to the entire nation. I do not think the general population appreciates how important software is to both our economic and military health." Dr. William Wulf has witnessed the software crisis from virtually every aspect. From his position with the National Science Foundation he oversees half of the federal government's funding for basic research on software. He is currently on leave from the University of Virginia where he serves as the AT&T Professor of Computer Engineering and Applied Science. He has seen the business side of it, having founded Tartan Laboratories, a software company.

But one doesn't need Dr. Wulf's impressive credentials to be aware of the explosion of software in our society. He has experienced most of the computer revolution as just another consumer. "Cameras, VCRs, and automobiles could not be sold without software. It's in wristwatches, telephones, and aircraft. Inventory control and automated order processing software are critical to economic methods of manufacturing. Software runs our elevators and provides environmental control in our office buildings. Software controls automatic teller machines, providing us twenty-four-hour banking. It controls the bar-code readers at our supermarkets. The list is virtually endless."

But, unfortunately, the proper development of software has failed to keep pace with the demands we are constantly placing upon it. As Wulf sums it up, "The production of software is both very expensive and unpredictable. It is often delivered over budget and behind schedule. When it is delivered, it is almost always incorrect."

Wulf shrugs. "I suppose one could say, 'So what? You know, everything is expensive and takes longer than we want, and so what is the difference?'"

The difference is that we live in an information society. Our economic and military vitality and often our own safety depends on the timely and accurate flow of information.

Sheryl Handler is president of Thinking Machines, a leading supercomputer maker. She stresses that computers are no longer of importance merely to computer scientists. "It is a technology that allows work to be done better in virtually every field," she points out. "Whether you are a scientist doing experiments, or a businessman making a decision, or a legislator trying to make good policy, having access to information in a timely way is the difference between being a leader, or being a follower."

The amount of information our society deals in is staggering. The Internal Revenue Service stores enough information on American taxpayers to fill one million volumes of the *Encyclopedia Britannica*. When an airline purchases a Boeing 747 jumbo jet, it receives papers on maintenance equal to the entire weight of the plane. As Dr. Handler notes, "Our society is changing, and if we do not have access to this information in a very powerful way, and before others do, then we will lose our lead in so many different areas."

The advancement of computer technology improves the quality of life for everyone, and produces effects that reach far beyond the narrow confines of the computer industry. Congressman Doug Wahlgren (D-Pa.), a member of the House Committee on Science, Space, and Technology, notes, "Many fields of science and technology require high-performance computing for advancements to be made. Examples include forecasting the weather, design of aircraft and

automobiles, recovery of petroleum resources, structural calculations for design of pharmaceuticals, and plasma dynamics for fusion
energy technology.''

The impact of software ripples through the entire economy. As
Justin Rattner, director of technology for Intel Corporation's Scientific Computers Division explains, ''We cannot overstate the importance that high-performance computing will play in the industrial
and commercial sectors as well as the aerospace and defense sectors.
You can look at virtually any industry—the automotive industry,
the aircraft industry, the oil industry, the transportation industry,
the semiconductor industry, materials, chemicals—all these industries will be highly, if not totally, dependent on high-performance
computing technology for their competitive edge.''

The computer revolution is without a doubt the fastest growth of
technology the world has ever seen. For less than one hundred
thousand dollars, one can now buy a supercomputer small enough
to fit on a desktop. That compact computer packs more computing
power than all the computers in the world did in 1959.

Computer developers love to point out that if the automotive
industry had improved at the same rate, a Rolls Royce would cost
ten dollars, get one hundred thousand miles per gallon, cruise at a
million miles an hour, and be the size of a cigarette pack.

But in stark contrast with the rapid development of computers,
proper development of software has failed miserably to keep pace.
Jacob Schwartz directs the Information Science and Technology
Office for the Defense Advanced Research Projects Agency
(DARPA) at the Pentagon. He has been part of the computer industry
as a university professor and industrial consultant. He notes, ''the
physical capabilities of computers have advanced steadily and with
enormous rapidity over the last twenty years. Software technology
has advanced, but not nearly with the same steadiness or to the
same extent. This has turned software into the great unswallowable
lump, so to speak, of the whole computer field.''

And that lump is growing, as the gap between computer hardware
and software performance widens. While computers are becoming

faster at a rate of more than seventy percent a year, software efficiency is growing at a comparative crawl: just four percent a year. While computers have improved a thousand times in the last twenty years, the improvement in software productivity has been merely ten times.

"That is a major problem," Schwartz says, "and one on which we are only now beginning to get a grip." Computers are growing ever more sophisticated, necessitating even more complex software programs to run them.

Tennessee Sen. Albert Gore agrees. "Modern computer programs can be incredibly complicated," he states. "It is taking more and more time and effort to build the new software needed for a new computer system. Ensuring that such programs are error-free is a daunting task, to say the least."

Our inability to produce reliable, safe software has its origins in some very basic problems. Writing software is an extremely complex undertaking. Students typically struggle to write software that is a few dozen lines long. An average software programmer can write perhaps six lines of code per hour. But software for large business systems or the space shuttle may be tens of millions of lines long, with a complexity that is trillions of times greater than simple, rudimentary programs.

"Large software systems are among the most complex creations of the human mind," points out the National Science Foundation's William Wulf. Many software systems "have exceeded the ability of humans to comprehend them."

It is not surprising, then, to learn that the basic problem is also complex. While software has grown increasingly complicated, we have failed as a society to develop the skills of basic research that could teach us how to properly develop software. "In some sense," Wulf relates, "we have known the solution for a long time."

The difficulty, Wulf explains, is that "knowing *how* to do it is still a deep intellectual problem. The proper foundation, even the appropriate mathematics, does not exist."

One source of the bottleneck is a lack of trained people in the software field. David Nagel, a top researcher with Apple Computers, complains about the shortage of highly qualified software developers. ''The best software,'' he notes, ''is written by a small fraction of the total number of computer scientists. Better methods of training computing professionals are needed to improve this ratio.''

D. Allen Bromley, Assistant to the President for Science and Technology, concurs. ''Perhaps most serious of all, is the fact that we face a daunting shortage of people trained in this field.''

And yet research and education will be the keys to finding solutions to the software crisis. Dr. Ken Kennedy directs Rice University's Computer and Information Technology Institute. He notes that it is from the schools that the solutions will eventually come. ''Education and basic research are what are going to give us the ideas and the people that we are going to need to meet the challenge. We do not yet have all the ideas we need to solve these problems. We can't just call up a national lab and say, 'Here is a pile of money—solve the software problem.' We need ideas and the way to get ideas is to grow a research infrastructure in the United States which will be strong and which will generate those ideas.''

But alarmingly, the critical need for those people and ideas comes at a time when interest in computer sciences and related fields is dropping. Software consultant Ralph Crafts, president of the Ada Software Alliance, a group of leading software makers seeking to promote the use of the Pentagon-developed Ada programming language, complains, ''For a variety of reasons, we have not been attracting the best students to science and engineering for some time.'' He cites the difficulty of providing graduate students with the most up-to-date equipment and technical staffs as the main reasons cited by students for leaving graduate study, choosing instead to enter the workforce and pursue the high salaries of the computer industry.

Presidential Science Advisor Bromley reacts with alarm when he notes, ''Not only do we have a shortage of people in the existing pipeline, but there is also a very depressing poll of freshmen entering the nation's universities and colleges.

"To find that interest in the physical sciences has decreased by a third in the last decade is bad. To find that interest in mathematics has decreased by a quarter in the last seven years is frightening. To find that interest in computer science has decreased by two-thirds in the last four years defies my understanding. It does emphasize a dreadfully serious problem."

The problem is even worse among women and minorities, where a potentially significant pool of talent is going largely untapped. Minorities make up nearly twenty percent of the American population. But they account for only one and a half percent of the new doctoral degrees awarded in mathematical sciences. And even that tiny figure is down slightly over the last ten years.

Women, making up more than half of the population, constitute only thirty percent of graduate degrees in computational sciences and engineering. Martin Massengale, chancellor of the University of Nebraska, blames a failure of our educational system to encourage women to explore fields of science. Massengale feels strongly there is a "shortage of scientists and engineers, and underrepresentation of women in these fields. Between the ages of nine and thirteen, girls begin to lose interest in science careers. They doubt their ability to perform well or they question the advisability of excelling in math and science."

Rice University's Ken Kennedy puts it more bluntly: "We are losing women. Women are not selecting careers in computational science and engineering. They are not going on to graduate degrees for some reason."

But even those men and women who do enter science careers find that the training to provide them with the proper background to be software developers often does not exist.

Current software education does not give the needed engineering background. Software consultant Ralph Crafts complains, "The lack of an engineering orientation, especially in the area of software technology, is one of the most debilitating factors in U.S. loss of market share in global markets [in computer technology]."

A House Science, Space, and Technology Committee study concluded, "Software engineering education has not changed much of late," the report notes. "The field still resembles a 'shoot-from-the-hip' industry. Most students are not taught an engineering environment, just coding. Training today is delivered in a manner promoting lack of attention to design in favor of immediate attempts to generate code."

The problem stems from a failure to develop software engineering as its own unique discipline. "Software engineering is not the same as computer science," Crafts steadfastly maintains. "Computer science constitutes a small fraction of the disciplines, methods, and technologies encompassed by software engineering."

Computers are the first device made by man which is so complex that it required development of its own area of scientific research. But that research has often ignored the issue of how to properly develop software for it. William Wulf points a finger at the schools, saying, "Our educational system is struggling to cope with this new discipline; it is nowhere near adequate—not at the graduate level, not in the elementary schools."

Carnegie-Mellon University professor and former NASA computer scientist Dr. Raj Reddy notes, "If you asked how many trained computer scientists [there are] with bachelors, masters, and Ph.Ds in software science, software engineering, and other related topics, you will find that there are a very small number. Most of us have come into this discipline from other disciplines."

"It is appalling," Crafts points out, "to note that currently there is not one single undergraduate degree program in software engineering in this entire nation."

A lack of demand is partly to blame. There are currently no education requirements that software developers have to meet. Dr. Reddy says, "But I think requiring trained people to be available to do all the software projects and giving preference to those aerospace industries who have a lot of trained people, would do a lot to cause training to happen. Right now there are not many trained software engineering professionals."

Once they begin writing software programs, there is also a lack of written standards and practices for developers to use. In stark contrast with other types of engineering, such as structural or architectural engineering, software development still tends to be more of a seat-of-the-pants art, rather than a disciplined science. Reddy maintains, "Handbooks of software engineering practice would be very helpful, just like all engineers have handbooks. We do not have such a thing. Software is an art, and we understand a very large part of this art. It is about time we begin to codify this knowledge and make it available."

It is often suggested that software programmers be required to be professionally certified, just like architects, doctors, lawyers, and engineers. Professional boards would set standards of training and knowledge that software engineers would have to meet. They would also establish standards and codes of practice for developing software. Software developers would be required to use established software development practices that have been developed to try and improve the reliability of software. There would also be requirements for continuing education on the latest software technologies.

Software certification already exists in Great Britain. Software engineers must be certified if they write software for the Defence Ministry. All military software that is safety-critical must be written to a standard known as Def Stan 0056, thereby ensuring the software is written using practices known for developing safe, reliable software. To oversee the standard, independent software safety assessors monitor every project.

The British government also sponsors a program that strongly encourages the use of accredited safety-compliance procedures in civilian software applications where safety is of paramount concern, such as medicine, aviation, and nuclear power.

But in the United States the mere mention of professional certification draws howls of protest from software developers. They are, by nature, fiercely independent souls, and largely resistant to what they see as unnecessary outside interference. Michael Odala, president of the Software Entrepreneurs Forum said in *Newsweek* (vol. 115, January 29, 1990), "If there was a law that said you couldn't write software without a license, most of our members would go find other work.''

As Robert Ulrickson, president of Logical Services, a company that designs computerized instruments, complained in *The Sciences* (vol. 29, September/October 1989), ''I'll fight them to the death. I don't want to be part of an economy that's run by the government.''

But objections to professional certification border on the absurd. Licensing is required for architects who design bridges, but not for people who write the software programs the architects use on their computers to come up with those designs. Many states require certification for beauty salon stylists and barbers; no such certification is needed for people to write software that is incorporated into medical devices and jetliners. One computer instructor recalled the instance of a former student who was hired to write a software program for a jet fighter, even though he did not know how to fly.

Dr. Joyce Little, vice chair of the Institute for Certification of Computer Professionals, says, ''People are going into programming with no background. All you have to do is hang out your shingle. It's pretty horrifying. If the industry does not police itself by means of voluntary certification, then there is going to be licensing. I am not necessarily for licensing. I see it as a necessary evil that is going to come.''

Beyond professional certification, whether or not the problems of software development can be solved is largely up to what the federal government does. As the world's largest software customer, buying $100 billion of software each year, the government is in a unique position of leverage to encourage drastic changes in the software industry. In addition, only the federal government has the resources to coordinate research and facilitate the new technologies that will be needed.

The government is already making some moves to remove the software logjam. The National Science Foundation has set up the Office of the Directorate of Computer and Information Science and Engineering, to coordinate and encourage basic software research. Two grants were awarded to Rutgers and Rice universities to study new software technologies. The NSF is also studying an initiative that would create formal standardized methods for laying out the specifications for software in its design stages.

Much hope rests in the hands of the Defense Advanced Research Projects Agency. DARPA already spends about $25 million a year on software research. As DARPA's director of Information Science and Technology, Jacob Schwartz thinks his office is in a unique position to help find answers to the software crisis.

"We are already in the process," Schwartz points out, "of launching a major new activity aimed directly at the general simplification of software design." DARPA is developing a new technology called prototyping. Workbench software systems are set up that show the user how the final software will work. An army general, a pilot, a business executive, or whomever the ultimate user will be, can come in and see the software in action before it is completed. Changes can then be suggested, and the software can be altered before it is produced in its final form. For example, a business executive can check to see whether the software is giving the information he or she needs, and in a form that they can use.

Prototyping offers the promise of producing software faster and cheaper, while also ensuring greater safety and reliability. TRW Corporation has already established a massive Rapid Prototyping Center. Using a large computer, TRW can construct a working prototype of a new software program within a few days, which will show prospective customers precisely how the finished product will work. Ironically, the solution to the software crisis may lie in the greater use of computers. Prototyping requires use of large amounts of computing power in order to set up demonstrations to show to users.

Another computer-driven potential solution to the software crisis comes from Japan, where they have developed a system called Daisys. An outgrowth of Japan's supercomputer research, it uses computers to design and build software programs. It allows users to define in their own terms and terminology precisely what it is they wish the software to do.

This system reduces miscommunication between the software user and the program developer, the latter role being assumed in this case by the computer. After evaluating the system, software consultant Jim Webber said, "It's really quite remarkable. It makes the user both the programmer and the system designer."

But development of these new products is being hindered by a lack of common computer languages. There currently exists a "Tower of Babel" that dilutes efforts to improve software. Dozens of incompatible software languages mean advances made on one computer system cannot be readily transferred for use on other computers.

It is here that leadership from the federal government could greatly improve the situation. Senator Gore readily recognizes the need for some standardization of software languages. "A lot of software," he explains, "is written to match the requirements of the hardware on which it is going to be run. Consequently, it becomes extremely tedious and almost impossible in many cases to translate that same program for use on a different piece of hardware with a different operating system. Many software developers and software users have pointed to the need for easy ways to transfer programs from one system to another, so that if you have a solution for one problem, it is available to all the users of advanced computing systems."

Software expert Ralph Crafts calls the current myriad of languages "one of the major contributors to the current software crisis."

But computer developers are already balking at the idea of computer language standardization. Samuel Fuller, vice president of research for Digital Equipment Corporation complains, "with too many standards, I believe, we will tie the hands of developers and inhibit innovation."

Senator Gore complains, "Computer designers tend to resist standards because the standards limit their ability to design faster and more innovative hardware and software by taking unique and novel approaches to solving problems."

Senator Gore has sponsored a bill called S.272 that would, among other far-reaching things, direct the National Institute of Standards and Technology to develop software standards, which would make it easier for software to be transferred from one computer system to the next. The bill would also achieve several other major objectives:

— Coordinate the government's software research, currently scattered among the Pentagon, NASA, the National Science Foundation, the Energy Department, and several other agencies.
— Establish a national high-speed computer data network, an electronic "interstate highway" that would shuttle information between computers in far-flung cities. "Increasing access of one team of researchers to another," says Gore, "and allowing the easy distribution of software would really supercharge the various research and development efforts throughout the United States."
— Direct the National Science Foundation to set up clearinghouses that would collect software and maintain software libraries. This would help researchers find the software they need for particular tasks.
— Provide funding to researchers to improve software they have already developed.

The emphasis on basic research is perhaps the most important aspect of the bill. Although the idea of fundamental research sounds terribly unappealing, and produces few *immediate* results, it is probably the single most important step that can be taken to improve the quality and safety of software for the long term.

The NSF's William Wulf rolls his eyes at the notion that this suggestion lacks excitement. "This is perhaps the least glamorous action that one could contemplate. But the most important action is to continue our commitment to long-term fundamental research. Such research is the source of ideas and steady progress. It is an investment in the future of our country. I've tried to make it one of my highest priorities, and we are starting to make some progress,

I think. I do not believe that we even have the appropriate mathematical basis for characterizing software, and lacking that, we will not develop the appropriate engineering base either. Much of the fundamental software problem will hinge on developing a better understanding of the theoretical aspects of the problem.''

John Poduska, chairman of Stardent Computers, a leading supercomputer maker, agrees. "Support for basic research and education is a crucial and fundamental element for long-term success.'' And yet it has long been neglected. A National Science Foundation panel notes, "Computer science research is underfunded in comparison to other scientific disciplines.''

Gore's bill would go a long way towards addressing that deficiency. It would spend nearly $2 billion on software research. "It is an ambitious plan," Gore admits. "This is important. There are incredible benefits to be gained in this field if we make the necessary investment.''

The Tennessee senator concedes funding for this effort will be hard to come by, "due to the budget deficits we face. But frankly, we cannot afford not to fund this effort. Budgets are tight, and it will not be easy to find the necessary funding. But this is new technology, and we need new money to develop it. Our economic competitiveness and our national security depend on it.''

Sen. Ernest Hollings agrees computer software is not exactly the hottest topic on Capitol Hill. He admits it is "an aspect of computer technology that often gets less attention than it should. You can't see software, and you can't touch it. It is much easier to talk about something tangible, like the hardware that makes up a computer. Yet software is critically important.''

John Rollwagen, chairman of Cray Research, notes the glaring disparities in funding priorities for economic development. "We invest millions of dollars in iron and steel, and very, very little in silicon," he says. "And yet, it is the silicon that provides the intelligence to the iron and steel, to the satellites, to the data-gathering systems we have.''

As Dr. Raj Reddy points out, the software industry contributes

$400 billion to the nation's gross national product each year. "It is an industry," he says, "which we can ill afford to ignore and leave to chance."

Gore's bill, S. 272, and its companion bill in the House of Representatives, H.R. 656, have picked up the support of the White House. But even if passed, it may be decades before the full benefits of the bill are realized.

Although the software crisis has been looming for years, it is imperative we get a handle on it now, before the gap between computer technology and the software that runs it grows even wider. Software engineers continue to be hard at work developing so-called "artificial intelligence" software. The software would learn and adapt itself to its environment, in much the same manner that humans learn from their mistakes and begin to recognize their surroundings.

But because AI, as it's called, by definition adapts itself as conditions change, its behavior becomes impossible to predict. The software programs will react according to what they have "learned," outside of what they have been specifically programmed to do. As congressional researcher James Paul states, "Adaptive intelligence in safety-critical systems may be beyond the capability of regulatory agencies to certify."

Software developers already speak of cars with electronic brains that will automatically avoid accidents. Sensors would detect obstacles and steer cars around them. Scientists estimate such sensors could eliminate eighty to ninety percent of all fatal accidents, yet cost less than ten percent of the total cost of the car. Prototype systems are already under development.

Translating telephones could be another major advancement. Japan has begun a $120 million project to develop telephones that will simultaneously translate a conversation into another language, such as Japanese or German, while you speak. The technology could be a great boon to the international business community.

Men and women of that community will soon carry computers in the palms of their hands that will be more powerful than the computers currently in their offices.

Other computer scientists speak freely about developing computers within five years that will be one hundred times faster than today's most advanced supercomputers. But these wonderful new advances depend on our ability to produce software that is up to the task.

Dr. Herbert Freeman, director of Rutgers University's Center for Computer Aids for Industrial Productivity, points out that software is the potential choke point for this. "The development of software has always lagged behind advances in hardware. And without high-performance software, the new computers will not realize their potential nor become cost-effective."

In fact, our current deficient level of software fails to even take full advantage of the computer technology we already possess. Experience at NASA has found that its software programs make use of barely twenty percent of the computing power available on its massive Cray-2 supercomputer. David Nagel of Apple Computers complains, "It is clear that we do not yet know how to build software systems that can take full advantage of these new parallel architectures. Computer software shapes up as the major bottleneck."

And while the United States flounders in its efforts to solve the software crisis, other nations are racing ahead to take advantage of that opportunity. Representative Wahlgren notes with alarm, "Our economic competitors in Europe and Japan are well aware of the importance of high-performance computing, and the United States, while still the world leader, is now in a race for dominance."

Dr. Sheryl Handler, of Thinking Machines, Inc., puts it more succinctly: "The magic question is, the world is going to change, but is America going to be in the position of leadership as it changes? If we are not, someone else will be."

"The United States is the world leader in the software industry, but we are not appreciably ahead of the rest of the world in basic software research," points out Dr. William Wulf. "Japan has clearly enunciated national software research goals, and has initiated well-developed government-industry-university programs. Europe has historically taken a more mathematically oriented, fundamental approach to software research. By means of large programs that team professionals from universities and industry, they are making significant strides towards applying these results.''

Japan is in the midst of a concentrated, five-year, $200 million project called SIGMA (Software Industrialized Generator and Maintenance Aids), which it hopes will vault it into the lead in worldwide software development. SIGMA teams up the government with 187 companies, including computer manufacturers and software companies, to develop new software technologies in a field where Japan still trails the United States.

A recent study by the Massachusetts Institute of Technology found that Japanese software has, on average, fewer than half as many errors as the average U.S. program.

The European Community has launched a similar project called ESPRIT, the European Strategic Program for Research and Development in Information Technology. ESPRIT encompasses some two hundred different projects, combining the massive resources of companies including Olivetti of Italy, Germany's powerful Siemens, GEC of Britain, and the French Groupe Bull. Lazlo Belady, vice president of the U.S.'s Microelectronics and Computer Technology Corporation, warns bluntly, ''The real software competitor is Europe, not Japan.'' ESPRIT is pursuing new methods of developing software that he says are aimed at ''giving more assurance that software does not contain errors and will perform as specified.''

ESPRIT is already producing dividends. A European Community project called Eureka is providing funding to a new facility that will use state-of-the-art technology to develop more reliable software.

Still and all, the United States is perhaps best equipped to solve the software crisis. The free-wheeling, independent nature of its software industry and university research generates the creative energy that can develop solutions. John Poduska, chairman of Stardent Computers, points out, ''The United States is in a unique position to exploit advantages in advanced computer software for science and engineering. We have a community of scientists who are willing to take professional risks. We have a community of entrepreneurs and a venture capital structure that are willing to support risky new efforts in software development.''

As Robert Paluck, chairman of CONVEX Computer Corporation, points out reassuringly, ''American companies continue to be the premier software development houses in the world.''

But, discouragingly, current government policies are hindering development of many promising new software technologies. Federal government computer procurement regulations are absurd. When the government purchases new software, they require that the software developers turn over to the government not only the software, but any new technologies or tools they developed to help them write the software.

Dr. William Wulf calls this simply ''disastrous'' for encouraging new software technology. ''Either contractors do not make the investment [in new technologies] and continue to use labor-intensive, non-modern, old-fashioned techniques, or else, if they have made the investment, they do not use their best tools and technology on government software, and consequently the government gets less high-quality software, it pays more for it, and it takes longer to develop than it should.''

Jacob Schwartz of DARPA shakes his head at the government's own software procurement regulations. ''The problem is inherently difficult,'' he says. ''In my belief, government policies made it considerably worse.''

Some in government recognize the need for changes in government regulations. Senator Gore's S.272 bill would alter the government's procurement regulations.

As Senator Gore explains, ''Computer technology is advancing at breakneck speed. Policymakers must be sure that government policies and research priorities do not lag too far behind. I am fond of quoting Yogi Berra, who once said, 'What we have here is an insurmountable opportunity.' ''

Presidential science advisor D. Allan Bromley sees a glimmer of hope. ''We still have the opportunity—a window of opportunity, if you will—to take a leadership position in the world on this one,'' he claims. But he also warns, "the window, I believe, is a relatively narrow one."

Dr. Raj Reddy agrees the time for action is running short. In written testimony before Congress he stated, "Most of the research proposed in S.272 is not only important but may be essential for maintaining the U.S. technological lead. *Indeed, it may be too little, too late!* (the emphasis is Reddy's)"

The current situation, he says, "calls for an aggressive industrial policy on the part of the government in an area of strategic interest in the country. A five- to ten-billion-dollar-a-year investment into research and development may be necessary."

Solutions will need to be far-reaching. They must include more basic research, improved educational facilities, streamlined government procurement, better training and standards for software developers, and more government funding. House investigator James Paul concludes, "No single action will magically solve the software woes. Rather, the solution will be found more in a sustained and difficult series of apparently mundane—but eventually far-reaching—changes in policies." But the need for action, he notes, grows every day.

Dr. William Wulf rests in his office at the National Science Foundation headquarters in Washington, D.C. It is something he rarely gets to do. Life is a whirlwind of meetings, lectures, and trips as he directs the flow of federal funding for basic software research.

His crusade to correct the problems of software will not pay off for years. He worries that people think the software crisis is simply a problem for the computer industry to worry about.

"That is wrong!" he steadfastly maintains. As a professor, scientist, and entrepreneur, he has watched computers creep out of the laboratories and into automobiles and airplanes. He witnessed the birth of the computer in the 1940s, and then watched in amazement as computers shrank from the size of a room to a desktop workstation.

"Software is crucial to all aspects of the country's vitality," he says. "You have to be able to produce software if you want to produce microwave ovens or VCRs or automobiles or virtually any consumer product as we go out into the next decade and the century. Do we want to sell automobiles?" he asks point-blank. "Do we want to sell airplanes? Software is a major problem."

But much more federal leadership and research funding will be needed. Even as he surveys the impressive work his agency is doing, Dr. Wulf is also clearly alarmed over the need for far more work. "It is time we recognize what is probably the obvious," he says. "And that is that it is a very hard problem. We have tried lots of solutions. Lots of ideas have been proposed and we have made progress, but we have not solved the problem."

The software crisis has been slow in developing. It will be equally painstaking to solve, requiring a concerted effort by government, industry, and educational institutions.

Wulf shakes his head. He concludes wearily, "We are not going to find a quick and easy solution."